W0079896

Lecture Notes in Economics and Mathematical Systems

675

For further volumes:
http://www.springer.com/series/300

Anna Maria Gil-Lafuente • Constantin Zopounidis

Editors

Decision Making and Knowledge Decision Support Systems

VIII International Conference of RACEF,
Barcelona, Spain, November 2013 and
International Conference MS 2013,
Chania Crete, Greece, November 2013

 Springer

Editors
Anna Maria Gil-Lafuente
Faculty of Economics and Business
University of Barcelona
Barcelona
Spain

Constantin Zopounidis
School of Production Engineering and
    Management
Technical University of Crete
Chania
Greece
and
Audencia Nantes School of Management
Nantes
France

Real Academia
de Ciencias Económicas y Financieras

ISSN 0075-8442   ISSN 2196-9957 (electronic)
ISBN 978-3-319-03906-0   ISBN 978-3-319-03907-7 (eBook)
DOI 10.1007/978-3-319-03907-7
Springer Cham Heidelberg New York Dordrecht London

Library of Congress Control Number: 2014956930

Printed on acid-free paper

Springer is part of Springer Science+Business Media (www.springer.com)

Preface

The Association for the Advancement of Modelling and Simulation Techniques in Enterprises (AMSE) and the Royal Academy of Economic and Financial Sciences of Spain (RACEF) are pleased to present the main results of the International Conference of RACEF, held in Barcelona (Spain), and the International Conference of Modelling and Simulation in Engineering, Economics and Management, held in Chania, Crete (Greece), 7–8 November 2013, through this Book of Proceedings. We would also like to thank Constantin Zopounidis from the AMSE for organizing the conference and Prof. Jaime Gil-Aluja, President of the Royal Academy of Economic and Financial Sciences of Spain, from RACEF.

MS'13 Chania is co-organized by the AMSE Association and the Technical University of Crete, Greece. It is co-supported by the Spanish Royal Academy of Financial and Economic Sciences and the Spanish Ministry of Education, Culture and Sports. It offers a unique opportunity for researchers and professionals to present and exchange ideas concerning modelling, simulation and related topics and shows how they can be implemented in the real world. In this edition of the International Conference, we want to give special attention to the emerging area of Computational Intelligence. In particular, we want to focus on the implementation of these techniques in the Economic Sciences, Culture and Sports. Thus, the title of this book is "*Decision Making and Knowledge Decision Support Systems*". It is a very broad research area that includes fuzzy set theory, neural networks, evolutionary computation, probabilistic reasoning and chaotic computing as particular research topics of this discipline. The growing importance of Decision Making in the Economic Sciences is obvious when looking at the complex world we are living in. Every year new ideas and products appear in the markets, making them very flexible and with strong unpredicted fluctuations. Therefore, in order to deal with our world in a proper way, we need to use models that are able to assess imprecision and uncertainty.

Barcelona, Spain
Chania, Greece

Anna M. Gil-Lafuente
Constantin Zopounidis

Acknowledgement to Reviewers

Special thanks to Anna Klimova and Víctor Alfaro and all the referees for their advice in the revision process of the papers submitted to the MS'13 Chania and the International Conference of the Spanish Royal Academy of Financial and Economic Sciences:

Yuriy P. Kondratenko
Emilio Vizuete
Sefa Boria

Contents

Globalisation and Crisis of Values

Lorenzo Gascon

Mr. Chairman,

Academicians, Professors, Ladies and Gentlemen,

The President of the Montenegrin Academy of Sciences and Arts, Professor Momir Djurovic, in his maiden speech, on the month of April 2012 in Barcelona, as new correspondent Academician for the Republic of Montenegro of the Royal Academy of Economic and Financial Sciences of Spain, stated that he, as with almost all the scientists in their late and mid careers, the interests are gradually shifting towards global issues.

Personally, when the President of the International Conference on Modelling and Simulation Professor Constantin Zopounidis invited me, I chose, in this line, to think aloud about 'Globalisation and crisis of values'. In a certain sense just as a way to complement/balance the high technical and scientific program to be developed by the bright list of participant Academicians and Professors.

Whoever tries to say something, nowadays, regarding globalisation, cannot refrain to face with the existing crisis, which embraces not only the economic and political world, but also the moral values on which the society should base its normal working.

The greek origin of the word crisis has its roots, in a great measure, in the worst illness that our world is suffering: the lack of the moral virtues. Especially, the cardinal: prudence, justice, fortitude and temperance.

Either these virtues recover their supremacy in every person—man or woman— or the society will, fatally, precipitate into the chaos.

Many consequences of globalisation are hosting alarm signals with regard to the present weakness of the economic governance.

L. Gascon (✉)
Royal Academy of Economics and Financial Sciences, Spain

Fondation Jean Monnet pour l'Europe, Lausanne, Switzerland
e-mail: ll.gascon@economistes.com

A.M. Gil-Lafuente and C. Zopounidis (eds.), *Decision Making and Knowledge Decision Support Systems*, Lecture Notes in Economics and Mathematical Systems 675, DOI 10.1007/978-3-319-03907-7_1, © Springer International Publishing Switzerland 2015

Years and years of unregulated and risky deals have exposed huge financial corporations to colossal losses. It is a framework of mismanagement. And often corruption.

And waste and misgovernment with budgets beyond control.

We are enduring deep changes in government, the markets, corporate behaviour and the handling of information. We strongly need guidelines to help and to set us on the right path.

Let's say that, nowadays Ethics is lacking almost everywhere. Ethics and Trust have to be restored.

Transparency and making the inner work visible has to be the way to all. But, pressure on governments and companies to clear their acts is making it difficult.

When a government or a company makes up or inflates its figures, everyone suffers: people, investors, financial markets and employees.

An indispensable condition in the human relationship is credibility. Without trying to offend any of the politicians of the global society where we are living, it is easy to observe how the personal and institutional authority decays whenever the citizen feels that his expectations have been frustrated.

And here, we have to come back to the crisis of values that we are enduring in what is called globalisation.

Michael Elliott in the 'World Economic Forum' held in Davos in 2009, said that the global economy offered a little attractive panorama.

One of its main manifestations is the crisis of Ethics. And the indifference towards any moral rule.

The target for most corporations is to maximize profits for those who have the control.

Here comes the criticism for the dominion of the economy by a few multinationals that, even dare, and actually are in a position to dictate the politicians what they have to do.

Ethics is lacking everywhere. Trust has to be restored in governments and corporations.

Executive compensation, out of control, is likely to fall back in an effort to create a sense of fairness.

Moreover, those who helped to create the problem are receiving huge bonuses with the losers or the government money... Here, we cannot forget the club of many unscrupulous consultants and auditors.

Victor Hugo thought: 'More powerful than armies is an idea whose time has come'. And we should add: only ideas can reverse the present trend.

The lack of ethics appears almost everywhere, but, notoriously, in the political field. For example, in the way that debt is treated. Let's stop a little in this issue, as a token.

What is a debt? Individuals and nations can pay for their purchases in three ways: 1) out of current earnings; 2) by drawing from past savings; 3) by going in debt.

The budget deficits are partially attributed to a man called John Maynard Keynes (the Keynesian economics). Keynes, as we know, one of the greatest and most famous economists of the twentieth century, argued that during a time of depression and unemployment, such as that experienced during the Great Depression, the government could intervene into private affairs and manipulate savings to offset the generally depressed and unemployed economy.

The way he proposed to accomplish this was by increasing 'national aggregate demand', by running a government budget deficit during depressed years. However Keynes never resolved how this outstanding deficit would be financed. Keynesian deficits 'had to be financed' and that was the unpleased catch. Through inflation?

Inflation is not a magic potion. Essentially, inflation is a silent tax eroding wealth from the wage earners.

Another example is how promises and commitment before voting in the polls have a harmful influence at mid and long term.

Friedrich Hayek stressed the symbiotic relationship between morality and freedom. He wrote that 'it is an old discovery that morals and moral values will only grow in an environment of freedom and that, in general, moral standards of people and classes are high only where they have long enjoyed freedom. Moral convictions are necessary for a free society'.

But Hayek has emphasized that concepts of 'social justice' and 'economic rights' are, sometimes, among those that are incompatible with freedom. Unfortunately the concepts of 'social justice' and 'economic rights' are so ill—defined it imposes no limits on the claims which can be made under these banners.

In practice, these concepts are likely to become a mere pretext for claims for privileges by special interests.

The result is that every single act of this kind will give rise to demands by others to be treated on the same principle. And these demands can be satisfied, only when the government will perform these tasks with public funds and always pending of the votes in the poll. With abstraction of the budget deficit that these policies will carry.

Pierre Mendes-France Prime Minister of the French Republic in the fifties, in his book Moi et la Republique (I and the Republic) says: 'Qu'est-ce que c'est la politique?' (What's politics?). And answers: 'La politique c'est l'argent des autres'. (Politics is the others' money).

Democracy means rule by the majority. And if those who are relatively poor are in a majority they should be in a position to benefit themselves by seeing it that the government adopts and follows a policy of transferring wealth from the relatively rich to the relatively poor.

Some people are beginning to conclude that our present standards of living, production and accomplishment have been reached as a result of social security, unemployment insurance, public housing, price controls, poverty and welfare programs, farm price supports, and aid programs to this and that...

Persisting economic instability is aggravating the financial situation. Many governments are suffering huge budget deficits that are draining the capital markets, sometimes boosting interest, and causing large trade imbalance rates.

From its very beginning, many economists have strenuously opposed all political efforts at redistribution. Many redistributionists nevertheless favour progressive taxation because they are more concerned about inequalities of income and wealth than the alleviation of poverty. This is why many redistributionists favour a floor beneath which no one should fall and a ceiling above which no one should be permitted to rise.

When government taxes some to give to others, it is not engaging in charity, but promoting legal plunder.

What is important? As Leonard Read said, 'principles are important'. Moral philosophy is important. And courage and an abiding faith in one's convictions are important.

Richard von Weizäcker, Ex-President of the Federal Republic of Germany for 10 years, in the occasion of receiving the 'Premi Internacional Catalunya 1995' (Catalonia 1995 International Award) said that 'we are indebted to the Athenian Solon for the first constitution of a democratic nature. He defined politics as the function of balancing the disparate interests of the citizens, and pointed to ethics as an objective of the constitution. Is ethics the real objective of the interest of the citizens as Solon believed it should be? The person who defends his own interests without reference to a clear ethical foundation has a deficient understanding of these interests; and in the long term, will end prejudicing them. We cannot separate politics from ethics, but neither can we separate ethics from politics'.

But, generally, politicians are not a different class of person from those who have chosen them for this role through election. Many of the disappointments which people have recently experienced regarding their politicians in all democracies have their origin in the observation that electors and politicians are of the same class; that is, they are equal.

The same human frailties are being denounced everywhere. The political world is fraught with problems of all kinds, with nepotism and favouritism. Political power is used for personal gain. Cases of corruption inflame the spirit of the people. The more the actions of politicians diverge from their earlier promises the more their credibility is reduced.

The State's lack of morality destroys the basis of the moral of their citizens.

Is the moral crisis doomed to become permanent? At long term, certainly not. Moral is inherent to the human conscience. Should it disappear we would confront a mutation of the species.

To Ethics happens the equivalent to the stature, the weight or the colour. One cannot live without them.

In 1848 the French economist Frédéric Bastiat wrote that 'he loses patience completely over claims that government spending programs can create jobs. Whenever the State opens a road, builds a palace, repairs streets, or digs a canal, it gives jobs to certain workers. That is what is seen. But it deprives certain other labourers of employment. That is what is not seen'.

Ludwig von Mises, the Austrian defender of human liberty and free enterprise, said that Government is evil. Although a necessary and indispensable evil. Government as such is not only not an evil but the most necessary and beneficial institution, as without it no lasting social cooperation could exist.

Men are born unequal and it is precisely their inequality that generates social cooperation and civilization. Equality under the law was not designed to correct the inexorable facts of the universe and to make natural inequality disappear.

Society is nothing but the combination of individuals for cooperative efforts.

Man has only one tool to fight error: reason. Man uses reason in order to choose between the incompatible satisfaction of conflicting desires.

The end of Science is to know reality. Obviously Science always is and must be rational, following reason.

Private ownership of the means of production (market economy or capitalism) and public ownership of the means of production (socialism or communism or planning) can never be confronted with one another, they cannot be mixed or combined.

Von Mises said that society is joint action and cooperation, in which each participant sees the other partner's success as a means for the attainment of his own.

A lot has been said about the symptoms of the historical crisis. Especially nowadays. The lack of engagement with the rules of values, that long ago were unquestionable, belongs so much to the symptoms of a crisis as to the loss of prestige of the established authority.

Every time that in History has been put in question the recognized authority—and that is also valid for the roman consuls as for the Pope at the end of Middle Ages as, also, for the king in the eve of the French Revolution—society sinks in a general crisis. Now, we should say, in a global crisis.

Right now, in the globalized world where we are living, it would be highly interesting to know where is the origin and the consequence. In other words, is the weakness from above (for example, in a democracy, the arrogance that can be observed during a long period of permanency in power) is what determines the uprising of the base or is the constant and increasingly critical aggression.

Of the base what slowly paralyses the exercise of the authority?

I guess that it is not possible to find a valid and final answer, because, as usual in History, doesn't exist a monocausality but a very complex system of different considerations of retrocession that it is very difficult to set in landmarks.

What is right is that in a good measure, have been discovered the irregular and deceitful practices.

Their consequences are on sight, in the stock exchanges, in the confidence of the investors, in the reduction of the consumption of the citizens when a substantial part of their savings have volatilised, and finally in what is called global economy.

A curious feature of the economic scene is the persistence with which commentators regard government financial affairs as being somehow different from everybody else's. Government's indebtness both in its annual form of deficit and in its perennial form of accumulated debt, is merely a subject for debate. Some would reduce it, while others hold that it could be increased to the general advantage. Whether the debts are accumulating in Madrid, or Athens, or Paris, no date is set for repayment.

It is in this vagueness, this lack of a power to grasp, that the trouble lies. The housewife knows how much there is to spend, and budgets accordingly. So does the businessman. So, within certain limits, does the corporate executive. The housewife's budget may provide for mortgage payments, the businessman's for repayment of long term debt, the corporate executive's for obligations to debt and equity. But each one represents both payment toward eventual discharge and provision for the same out of earnings. Not so with governments. Whatever payments they may make are not to discharge the debt but to pay interest on what was borrowed before plus the interest on what has been borrowed lately. That the money to pay interest may also be borrowed, until the borrowings compound one another like boxes in a Chinese puzzle, goes unremarked. This year's deficit commands the headlines. Yesterday's and tomorrow's fend for themselves. Participants in the debate who opt for reducing the deficit and, eventually, the debt, declare that either spending must be cut or taxes raised. Or, both. And this is unfair, unethical.

Likening the national budget to other budgets, they offer like treatment, excess of expenditure over revenue calls for expending less or earning more. Many believe the tacit assumption that it is the duty of the State to deal with all the evils and secure all benefits. Increasing power of a growing administrative organisation is accompanied by decreasing power of the rest of society to resist its further growth and control. The people, at large, led to look on benefits received through public agencies as gratis benefits, have their hopes continually excited by the prospect of more, and more.

It is safe to assume that a majority of citizens, having given the matter a moment's thought, will conclude that none of the benefits can be gratis. The public agencies' staffs must be paid, the facilities which they supply the benefits must be built and maintained, the equipment they use must be manufactured and paid for. All these costs are a charge on the public purse. Yet, any suggestion that the services be priced, and that the citizens who use them should pay the prices, is said to be politically unacceptable.

Ideally, bringing the reality of price to those services would be debated by political leaders. But self-interest prevents it. The fear of defeat at the polls is more pressing than the debt that stems from the policies. They are in what psychologists call a social trap. Like drug addicts who know that the addiction is harmful, they seek temporary relief in exchange of lasting damage. Like those other addicts, they need outside help. It lies with the financial community whose spokesmen are most prominent in criticising the debt. Not that the nation's leading bankers and investment dealers are impartial.

Financing the debt constitutes a sizable part of their business. The trap they are in is similar to that of the politicians, with this difference: that of the financiers have the means to get out of it with advantage of themselves. The debt crisis presents them with the opportunity to assist governments in applying the same remedies that businesses were forced to adopt. Selling tangible assets, selling equity, and withdrawing from unprofitable activities.

Pope Benedict XVI in his Encyclical Caritas in Veritate published in July 2009 says that 'in face of the crisis that the globalised economy is suffering, the Church doesn't nave technical proposals to afford but she does have ethical solutions'.

The whole world is, right now, a closed economy.

The financial waves in a global world are hitting up the last corner. Nowhere is immune.

Non-ethical bankruptcies have been derailing the global markets.

'Greed is good'. 'He who has the gold makes the rule'.

But, never in the planet there have been so many people literate and with a mobile.

The planetisation of the scientific-technical revolution should enable us to live better and with dignity.

Without technology we would not be viable. We use glasses, we write with a ballpoint pen, we drive cars, we utilise mobile phones, etc. Everything that is human implies technology. Technology makes us humans.

In any case, high technology has brought the present globalisation.

The future will probably be defined by science democratisation.

Like a GPS instrument we are going towards a future of personal plenitude and planetarian equilibrium. We should have to read this human GPS as Global, Personal and Social. It could orientate us through this new world in full transformation.

All the social, economic and cultural structures that make our present world unsustainable have been born and grown in the human mind. And, therefore is in the mind where we can transform them.

Like the UNESCO by-rules declare in its article Number one, 'Since wars begin in the minds of men, it is in the mind of men that the defences of peace must be constructed'.

The values are the intangible rails which guide our actions.

Though it exists a full discipline dedicated to its study—axiology—, ethic values are not so easy to be defined. The philosopher Dr. Terricabras says that values are 'the principles, ideas, relationships and experiences which we believe that give courage, strength and grace to our lives'.

Among the traditionally recognized values (and not necessarily practised) we find the tolerance, equality, solidarity and liberty. Nothing to see with the stock exchange values. The French poet Paul Valéry ironised saying that 'there is a value named spirit like there are values called oil, corn and gold'.

In The wealth of Nations (1776), Adam Smith already pointed out this paradox: 'nothing is more useful than water, but it will purchase scarce anything; scarce anything can be had in exchange for it. A diamond, on the contrary, has scarce any value in use, but a very great quantity of other goods may frequently be had in exchange for it'.

This paradox has very much to see with the unsustainability of the contemporary economy.

Like the stock exchange values, the quotation of the ethic values also fluctuates. And with rapidity in times of uncertainty as in the present conjuncture. Which values should have to quote on a rise in our culture and in our imaginary to reach a society fully human?

The financial crisis initiated in 2008 has shown the excesses and weaknesses of the unlimited globalisation against which John Maynard Keynes had already warned 80 years ago: 'the ideas, the acknowledgement, the science, the hospitality, the travels: all these are, by nature, international. But, let's make the goods to be produced as near our homes as possible, reasonable and convenient, and specially let's make the finances to be under control'.

Among a so magnificent diversity of cultures and ways of living, we are a unique human family and a single earthly community with a common fate.

We have to unite ourselves to create a sustainable global society founded in the human respect towards nature, the universal human rights, the economic justice and a peace culture.

Summing up, the global situation is that the dominant patterns of production and consumption are causing an ambient devastation, exhaustion of resources and a massive extinction of species. Communities are being destroyed. The benefits of development resources are not equitably shared and the gap between rich and poor tend to widen.

The lack of justice, poverty, and the violent conflicts are expanding and causing great sufferance.

Also, an unprecedented growth of the human population has burdened the ecological and social systems. The foundations of the global security are threatened. These tendencies are dangerous but not inevitable.

Fundamental changes in our values, institutions and way of living are necessary. We must take the decision to live according to a sense of universal responsibility.

The contemporary identity tries to fill its void with the consumption of material goods. Consumerism stimulates narcissism, doing that each individual depends on the others to validate its self-esteem.

Ingelhardt makes use of a scale of materialism—post-materialism based in the answers to a battery of questions:

- To maintain the order in the country.
- To grant the people with more opportunities to participate in the main political decisions.
- To fight against the prices increase.
- To protect the freedom of expression.
- To maintain a high economic growth level.
- To try that our cities and landscape be nicer.
- To maintain a stable economy.
- To get a less impersonal and a more human society.
- To fight against delinquency.
- To progress towards a society where ideas would be more important than money.

One of the ideological changes consists on looking for a better quality of life instead of a higher level of living.

On the last 27th July, in Rio de Janeiro, Pope Francis I called for the rehabilitation of politics and a permanent dialogue to smoothen conflicts and to find solutions.

He emphasized the social responsibility. He said that we are responsible in the forming of the new generations, competent in economics and politics and strong in the ethical values. He stated that sin can be forgiven, but corruption must not be forgiven. The corrupt feels himself as a winner. Corruption instead of being forgiven has to be hailed.

On the past 15th May he had said that we are suffering a tentacular corruption and a selfish fiscal evasion at world scale. The will-power and possession—will have become unlimited.

Corruption is a trap, it is a lie. Corruption brakes progress in any society and harms liberty. It is a call to improve the lives of everybody through honesty and well done work.

Just to conclude, let's remind that the Oxford Dictionary defines Ethics as 'the rules of conduct recognised as appropriate to a particular profession or area of life'.

And the Moral as 'what is concerned with goodness or badness of human character or behaviour, or with the distinction between right or wrong'.

Models Estimation of National Social and Human Capital Qualities

Gorkhmaz Imanov and Rovshan Akbarov

Abstract Present paper studies models of the estimation qualities of national social and human capital. Input parameters of the social capital model are social-economic security, social cohesion, social inclusion and social empowerment. Model of the human capital describes by quality of education, health care, culture level of society, scientific and technical journal articles, innovations, patent applications field, labor productivity. Problem corresponding to the model is solved by proposed fuzzy method.

Keywords Quality of social capital • Quality of national human capital • Socialeconomic security • Social cohesion • Social inclusion • Social empowerment • Fuzzy methods

1 Introduction

Social and human capital plays significant role in economic development. Social and human capital influences economic growth not only directly but also through interaction with each-other. Social and human capital enables individuals, communities and firms to cope with the demands of rapid changes. Estimation of social and human capital quality is almost impossible by using traditional methods. We do attempt to estimate social and human capital quality by using fuzzy approach in this paper. Paper mainly consists of two parts—in the first part were investigate fuzzy model of social capital, in the second—fuzzy model for estimation of national

G. Imanov (✉)
Institute of Cybernetics, Azerbaijan National Academy of Sciences, B. Vahabzadeh, 9, Baku 1141, Azerbaijan
e-mail: korkmazi2000@gmail.com

R. Akbarov
National Aviation Academy of Azerbaijan, Bina, 25-th km, Baku 1045, Azerbaijan
e-mail: rovshanakperov@yahoo.com

A.M. Gil-Lafuente and C. Zopounidis (eds.), *Decision Making and Knowledge Decision Support Systems*, Lecture Notes in Economics and Mathematical Systems 675, DOI 10.1007/978-3-319-03907-7_2, © Springer International Publishing Switzerland 2015

human capital quality. Problems corresponding to the models are solved applying fuzzy method defining quality indices. World Development and Azerbaijan Indicators were used in the process of solution of the problems.

2 Fuzzy Model Estimation of Social Capital

In order to define social capital were used indicators social quality. The theory of social quality has been offered by U. Beck, V. Maesen, L. Thomese and A. Walker [1–3]. Social quality represents degree of participation of citizens in the social and economic life of a society in which their well-being and individual potential raises. Social quality defined on the base indicators of the four conditional factors—social-economic security, social cohesion, social inclusion, social empowerment.

Social-economic security indicators are following: number of square meters per household member (NSM); proportion of population living in houses with lack of functioning basic amenities (PPL); proportion of people covered by compulsory/voluntary health insurance (PHI); number of medical doctors per 10,000 inhabitants (MED); length of notice before termination of labor contract (LNT); proportion of employed workforce with temporary, non-permanent, job contract (PET); proportion of workface that is illegal (PWI); number of fatal cases (NFC); number of nonfatal cases (NNC); number of hours a full-time employee typically works a week (NHE); proportion of pupils leaving education without finishing compulsory education (PLE); study fees in school as proportion of national mean net wage (SFS); study fees in high school as proportion of national mean net wage (SFH); proportion of students who, within a year of leaving school, are able to find employment (PSE); people affected by criminal offences per 10,000 inhabitants (CRI); ecocivilization index (ECC). Quality of socioeconomic security index (SESI) is the output indicator.

Social cohesion indicators are following: Extent to which most people can be trusted (TRU); Trust to authorities (TRA); Trust to religion (TRR); Number of cases being referred to European Court of Law (ECO); Respect for parents (IFA); Blood donation (%) (BLO); Multiculturalism (tolerance) (TOL); Willingness to pay more taxes if you were sure that it would improve the situation of the poor (WMT); Help elders (VOL); Membership (active or inactive) of political, voluntary, charitable organizations or sport clubs (MVO); Frequency of contact with friends and colleagues (CWF); Sense of national pride (NAP). Quality of social cohesion index (SCOI) is the output indicator.

Social inclusion indicators are following: proportion having right to vote in local elections (POV) and proportions exercising it (PPV); proportion with right to a public pension (PEN); proportion of ethnic minority groups elected or appointed to parliament, boards of private companies an foundations (ETH); proportion of women elected or appointed to parliament, boards of private companies and foundations (WPA); long-term unemployment (12+ month) (LTU); proportion of population with entitlement to and using public primary health care (PPH);

proportion homeless, sleeping rough (HLP); average waiting time for social housing (WAI); school participation rates and higher education participation rates (HED); proportion of people in need received care services (PPN); density of public transport system and road density (TRD); number of public sport facilities per 10,000 inhabitants (NPS); number of public and private civic and cultural facilities (e.g. cinema, theatre, concerts) per 10,000 inhabitants (NPC); duration of contact with relatives (cohabitating and non-cohabitating (PRC). Quality of social inclusion index (SIQI) is the output indicator.

Social empowerment indicators are following: extend to which social mobility is knowledge-based (SOM); percentage of population literate and numerate (PLN); availability of free media (FME); percentage of labor force that is member of a trade union (TRU); percentage of labor force covered by a collective agreement (LCA); percentage of employed labor force receiving work-based training (TRA); index of democracy (DEM); percentage of organizations/institutions with work councils (WCC); percentage of the national and local public budget that is reserved for voluntary, non-for-profit citizenship initiatives (CIL); proportion of national budget allocated to all cultural activities (CUL); percent expenses of national and local budgets devoted to disabled people (DIL). Quality of social empowerment index (SEQI) is the output indicator. Indicators of conditional factors of social quality were adopted from [3].

In order to estimation indices of factors of social quality were proposed method, which is based on L. Zadeh's composite rules of inference [4] and consist of the following steps:

- Development of a table describing parameters of the model on the basis of information obtained from international organizations and experts. In the first column of the table shows the input parameters of the model, and in the following columns—terms and their intervals. The last column specifies crisp meaning of input parameters for a fixed period.
- Definition of membership degrees of the crisp meaning of the input parameters to the relevant terms.
- Determination of the minimum degree of membership to the corresponding term of input parameters, i.e. $\max_i \left(\min_j \mu_{ij} \right)$.
- Determination of the maximum of the minimum values of the degrees of membership to the corresponding term, i.e. $\max_i \left(\min_j \mu_{ij} \right)$.

The obtained value will reflect the quality of the social factor.

The proposed methodology is tested on the basis of information on quality parameters of the model of socio-economic security (Table 1). The source materials are obtained from the international socio-economic organizations and the expert opinion data. Information on the socio-economic security of Azerbaijan in 2010, which is given in the last column of Table 1.

Table 1 Parameters of the model of social-economic security

Input variable	Terms and its intervals				Azerbaijan
NSM	L	M	H	VH	L
	0–15	14–20	18–30	28–70	12,6
PPL	L	M	H	VH	M
	0,5–0,25	0,24–0,15	0,14–0,1	0,09–0	0,15
PHI	L	M	H	VH	
	0–10	9–21	20–60	59–100	
MED	L	M	H	VH	H
	0–300	299–350	300–400	370–600	36,8
LNT	L	M	H	VH	H
	1–31	30–51	50–61	60–90	60
PET	L	M	H	VH	L
	100–50	49–20	19–10	9–1	68
PWI	L	M	H	VH	VH
	0,5–0,2	0,19–0,14	0,13–0,09	0,18–0	0,002
NFC	L	M	H	VH	VH
	10–8	7–5	4–2	1–0	0,00128
NNC	L	M	H	VH	VH
	10–8	7–5	4–2	1–0	0,00172
NHE	L	M	H	VH	M
	50–44	43–39	38–36	35–20	42
PLE	L	M	H	VH	M
	50–20	18–9	8–7	6–0	10
SFS	L	M	H	VH	M
	6–3	2,9–2	1–0,5	0,4–0	2,8
SFH	L	M	H	VH	L
	7–3	2,9–2	1–0,5	0,4–0	6
PSE	L	M	H	VH	VH
	0–5	4–10	9–20	19–100	30
CRI	L	M	H	VH	VH
	180–80	79–50	49–20	19–0	13,5
ECC	L	M	H	VH	H
	0–0,2	0,19–0,5	0,49–0,7	0,7–1	0,632

To calculate the quality of the social factors the following terms are used: low (L), medium (M), high (H) and very high (VH), which are scaled in the interval [0, 1].

In order to define linguistic variables intervals following calculations have been used, given in [5]. The second stage we have determined the degree of membership of national indicators of socio-economic security to the appropriate term. In determining the degrees of membership, we have used triangular membership functions.

At the task level membership of 16 indicators of the terms is as follows:

Low (L)	Mean (M)	High (H)	Very high (VH)
$\mu_{NSM} = 0{,}32$	$\mu_{PPL} = 0{,}25$	$\mu_{MED} = 0{,}64$	$\mu_{PWI} = 0{,}05$
$\mu_{PHI} = 0{,}4$	$\mu_{NHE} = 0{,}66$	$\mu_{LNT} = 0{,}18$	$\mu_{NFC} = 0{,}003$
$\mu_{PET} = 0{,}72$	$\mu_{PLE} = 0{,}22$	$\mu_{ECC} = 0{,}65$	$\mu_{NNC} = 0{,}003$
$\mu_{SFH} = 0{,}5$	$\mu_{SFS} = 0$		$\mu_{PSE} = 0{,}27$
			$\mu_{CRI} = 0{,}58$
min: 0,32	min: 0	min: 0,18	min: 0,003

Among the minimum values the maximum value is determined, which is equal to 0.32. This value corresponds to the term—"low", thus defining quality index of socio-economic security: SESI as low. Likewise, the indices of the quality of social inclusion—SIQI is 0.86 (high), the index of the quality of social empowerment— SEQI is 0.9 (high), the index of the quality of social cohesion—SCQI is 1 (moderate). Estimated meanings of the conditional factors social quality give us possibility estimate social capital quality. In order to estimation social capital quality index (SCQ), we used fuzzy union operation, i.e.

$$\mu_{SCQ} = \max\left(\mu_{SESI}, \mu_{SIQI}, \mu_{SEQI}, \mu_{SCQI}\right) = \max(0.32; 0.86; 0.91) = 1(\text{moderate}).$$

The proposed model for estimation of social quality gives us possibility to define human capital quality.

3 Fuzzy Model Estimation Quality of National Human Capital

Human capital are one of the main factors, which provide development of the socioeconomic system. Fundamental concept of the human capital theory were founded by American economists, Nobel Prize Laureates, Shultz [6] and Becker [7]. According to definition of Organization Economic Cooperation and Development (OECD) experts human capital are "the knowledge, skills, competencies and attributes embodied in individuals that facilitate the creation of personal, social and economic well-being" [8].

The conventional standard to measure human capital stock has been largely categorized into three parts [9]:

– School enrollments rates, scholastic attainments, adult literacy, and average years of schooling are the examples of output-based approach;
– Cost based approach is based on calculating costs paid for obtaining knowledge;
– Income-based approach is closely linked to each individual's benefits obtained by education and training investment.

Professor Kwon [9] refers human capital is difficult to identify and measure directly. So many researchers have used indirect measures. Concept human capital needs to consider both of monetary and non- monetary characteristic. Human capital is closely linked to social capital. It is necessary to analyze the result of human capital measurement within the socio-cultural framework of a society and all levels such as individual, organization, nation.

Human capital on the national level express intellectual potential of society.

In this paper investigated problem of estimation quality national level of human capital. With this purpose as recourses of human capital taken following elements:

1. Education quality–QUE;
2. Level of the health care development–QHC;
3. Cultural level of society–QSC;
4. Innovation index–INO;
5. Quantity of patent–PAT;
6. Quantity of articles–ART;
7. Labor productivity–PRO;
8. Socioeconomic security index–SEQ;
9. Social cohesion index–SCI;
10. Social inclusion index–SII;
11. Social empowerment index–SEI.

Estimation parameters choose–national human capital quality index.–NHCQI.

4 Formation Parameters of Fuzzy Model

Linguistic parameters of the level education, health care, culture of society were define by scaling these parameters in interval [0–10]. As factices meaning were used results of expert opinion of specialist, which deal in this sphere.

In order to define quality of science resident patent applications index, scientific and technical journal articles, patent application per million populations, innovation index were used. For this purpose corresponding world development indicators [10] and component knowledge economy applied [12]. Labor productivity indicators borrowing from [11]. Indicators socioeconomic security index, social cohesion index, social inclusion index, social empowerment index.

In order to definition linguistic variables intervals were following calculations used [14]:

(1) Sort the values in the current dataset in ascending order $(x_1, \ldots, x_n)$;
(2) Compute the average distance between any two consecutive values in the sorted dataset and the corresponding standard deviation $(\Delta_1, \ldots, \Delta_{n-1})$;
(3) Compute the revised average distance between two remaining consecutive values in the sorted dataset—$AD = \frac{1}{n-1} \sum_{i=1}^{n-1} |x_i - x_{i+1}|$ and standard deviation

$$\sigma = \sqrt{\frac{1}{n-1} \sum_{i=1}^{n-1} (\Delta_i - AD)^2}$$ for sorted dataset—Δ_i

(4) iDefine the universe of discourse—U and its scope—R:
   $U = [\min - AD, \max + AD] = [LB, UB]$;
   $R = UB - LB$

5 Calculating Method of the National Human Capital Quality Index

Using the information of Azerbaijan in 2010, which is given in the last column of Table 2, we have by the above methodology, the index is computed. To calculate the quality of the national human capital factors used in the following terms: very low (VL), low (L), medium (M), high (H) and very high (VH), which are scaled in the interval [0, 1].

The second stage determined the degree of membership of national indicators of human capital to the appropriate term. In determining the degrees of membership, we used triangular membership functions. At the task level membership of 11 indicators of the terms is as follows:

VL	L	M	H	VH
$\mu_{PAT} = 0.09$	$\mu_{QUE} = 1.00$	$\mu_{QHC} = 0.4$	$\mu_{SEQ} = 0.70$	$\mu_{SCI} = 0.26$
$\mu_{ART} = 0.14$	$\mu_{INO} = 0.90$	$\mu_{QSC} = 1.00$		$\mu_{SII} = 0.87$
$\mu_{PRO} = 0.95$		$\mu_{SEI} = 0.26$		
min : 0.09	min : 0.9	min : 0.26	min : 0.7	min : 0.26

Among minimum values determined the maximum, which is equal to 0.9. This value corresponds to the term—"low". Thus defined quality index of national human capital: NHCQI = low.

Table 2 Parameters of fuzzy model of definition national human capital quality

Input variables	Linguistic meaning and intervals input variables					Azerbaijan
1. QUE	VL 0–2	L 1.5–4	M 3.5–6	H 5.5–8	VH 7.5–10	L 2.75
2. QHC	VL 0–2	L 1.5–4	M 3.5–6	H 5.5–8	VH 7.5–10	M 4.00
3. QCS	VL 0–2	L 1.5–4	M 3.5–6	H 5.5–8	VH 7.5–10	L 4.75
4. INO	VL 0–2	L 1.9–4	M 3.9–6	H 5.9–8	VH 7.9–10	L 3.05
5. PAT	VL 0–543	L 534–1095	M 1076–1628	H 1619–2170	VH 2104–2800	VL 24,4
6. ART	VL 0–238	L 227–484	M 473–730	H 719–976	VH 965–1250	VL 17
7. PRO	VL 265–28734	L 26025–55080	M 52371–86844	H 84135–115599	VH 112890–142245	VL 20842
8. SEI	VL 0–0.2	L 0.17–0.4	M 0.37–0.6	H 0.57–0.8	VH 0.77–1	M 0.4
9. SCI	VL 0–0.2	L 0.17–0.4	M 0.37–0.6	H 0.57–0.8	VH 0.77–1	VH 0.885
10. SII	VL 0–0.2	L 0.17–0.4	M 0.37–0.6	H 0.57–0.8	VH 0.77–1	VH 0.90
11. SEQ	VL 0–0.2	L 0.17–0.4	M 0.37–0.6	H 0.57–0.8	VH 0.77–1	H 0.650
NHCQI	VL 0–0.2	L 0.17–0.4	M 0.37–0.6	H 0.57–0.8	VH 0.77–1	

6   Conclusion

Proposed approach to estimation of social and national human capital quality gives possibility to define country's level compared to the world development indices. It is necessary to further develop this idea on the individual and organizational level for the forthcoming investigation of the social and human capital quality.

References

1. W. Beck, L. van der Maesen, A. Walker (eds.), *The social quality of Europe* (The Hague, Kluwar Law International, 1997), pp. 263–296
2. W. Beck, L. van der Maesen, F. Thomese, A. Walker (eds.), *Social quality: A vision for Europe* (The Hague, Kluwar Law International, 2001), pp. 307–360
3. I. van der Maeson, A. Walker, M. Keizer, *European Network Indicators of Social Quality – ENIQ – "Social Quality*, Final Report, European Foundation on Social Quality, April 2005, 105 p
4. L. Zadeh, Outline of a new approach to the analysis of complex systems and decision processes. IEEE Trans. Syst. Man Cybernet. **3**, 28–44 (1973)
5. J. Poulsen, *Fuzzy Time Series Forecasting* (Aalborg University, Esbjerg, 2009), p. 67
6. T. Schults, Investment in human capital. Am. Econ. Rev. **51**(1), 1–17 (1961)
7. G. Becker, Investment in human capital: A theoretical analysis. J. Polit. Econ. **70**(2), 9–44 (1962)
8. T. Healy, S. Cote, *The Well-Being of Nations, The Role of Human and Social Capital* (OECD, Paris, 2001), 118 p
9. D. Kwon, *The 3rd OECD World Forum on "Statistics, Knowledge and Policy" Charting Progress, Building Visions, Improving Life* (2009), p. 15
10. *World Development Indicators* (The International Bank for Reconstruction and Development/ World Bank, 2012), 430 p
11. *Knowledge for Development* (K4D), *World Bank* (2012), http://info.worldbank.org/etools/ kam2/KAM_page5.asp
12. The Conference Board Total Economy Database™ (2013), http://www.conference-board.org/ data/economydatabase

Risk Management in Sports Sponsorship

Ramón Poch Torres

Abstract Sports sponsorship, a new form of corporate marketing, has become a major source of funding for sports teams and events.

As an entrepreneurial activity, sports sponsorship entails a degree of risk for the company which, like all enterprise risks, must be managed and controlled using effective, efficient programmes that analyse and assess the potential impact on the company's finances and reputation, thereby allowing the sponsor to take steps to avoid or mitigate such risks.

Keywords Risk management • Risk in sponsorship • Sports sponsorship • Sport marketing • Sports teams and events

1 Sports Sponsorship

Sponsorship in sports is a tool that provides funding for athletic organisations while offering companies a new way to advertise and raise brand awareness. It has become a highly developed field thanks to the symbiosis between sport and the business world based on the communicative power of sport, which has a healthy, appealing image and occupies many hours of television. In other words, it is an ecosystem of value exchange in which one cannot live without the other.

The official definition of sports sponsorship Ibáñez [5] is an investment Campos [1], whether monetary or in kind, in an activity, person or event in exchange for access to commercial potential to be used to the investor's advantage, although it can be defined in other ways depending on the type, content and specific field of the sponsorship and consideration arrangements.

R. Poch Torres (✉)
Academician of the Royal Academy of Economics and Financial Sciences of Spain,
University of Barcelona, Barcelona, Spain
e-mail: POCHTORRES@telefonica.net

A.M. Gil-Lafuente and C. Zopounidis (eds.), *Decision Making and Knowledge Decision Support Systems*, Lecture Notes in Economics and Mathematical Systems 675, DOI 10.1007/978-3-319-03907-7_3, © Springer International Publishing Switzerland 2015

Athletic activities appeal to sponsors because, as a general rule, sport is associated with quality of life, aesthetics, passion, emotions, etc. However, it is important for sponsors to remember that these qualities may inverted in an instant (giving no time to react) and become entirely different, negative messages that force sponsors to change their communications strategies, depersonalise their support and diversify their sponsorship activities if these are too closely linked to competitive success.

Sports sponsorship has boomed due to changes Mandado [6] in advertising laws and the evolution of leisure culture in the Western world: leisure makes new demands on the individual's personal life, and sponsorship attempts to overcome the consumer's resistance to the constant barrage of advertising messages in conventional media. The global sponsorship industry is valued at over 50 billion USD per year, with an annual cumulative growth rate of 9 %. Sports sponsorships account for more than 80 % of this market.

One of the greatest advantages of sponsorship is the ease of evaluating visibility, audience, attendance, brand awareness and other factors, and the interactivity that associates certain attributes of sport with the sponsor's brand—in other words, using it as a metaphor for values that the company/brand wishes to convey, such as universality, teamwork, healthy living, hard work or optimism. Sponsorship is a wide-ranging field that encompasses advertising, public relations, promotion, etc.

Traditionally, sponsorship resources have been funnelled into sports with mass appeal, as the extensive media attention they receive guarantees a large audience. However, the glut of sponsorships in high-profile sports have led some companies to look for more profitable options, such as sponsoring sports with a more limited following, which allow the company to customise the image it wants to project and achieve high brand awareness.

Another new form of sports sponsorship De Andrés [2] that is rapidly gaining ground today is naming rights, whereby a sponsor acquires the right to have its brand associated with the name of an athlete, team or stadium, often with such success that the brand becomes inseparably associated with the object of sponsorship. A case in point is the Arsenal Football Club's Emirates Stadium, where Emirates Airline renewed its sponsorship agreement in 2012 for 150 million pounds sterling.

2 Social Networks in Sport

Social networks have changed the lives of millions of people around the world. They have also become a powerful advertising tool, giving rise to new sports sponsorship strategies on Twitter, Tuenti, Facebook and sport TV, digital media and events channels. Sponsorship in the field of social networks has become incredibly sophisticated; for example, a type of contract has been invented whereby a celebrity agrees to post a single "tweet" at a certain time and on a specific date.

3   Sponsors

The effectiveness of sponsorship is directly proportional to the level of support provided by the sponsor, although there is an unwritten golden rule about the high rate of return on sports sponsorship which says that for every 4 euros invested a sponsor will make 5.

Based on this understanding, many successful companies Davara Rodriguez [3] have invested the bulk of their advertising budget in sports sponsorship, such as Nike, Coca-Cola and particularly Red Bull, which sponsors spectacular, high-risk events in an attempt to associate its brand with youth and energy. By way of example we might mention Felix Baumgartner's famous space jump in October 2012, which cost Red Bull 30 million euros but immediately caused its sales to skyrocket and increased its brand value to 14 billion euros, according to the European Brand Institute.

Examples of sports sponsorship in Spain include Banco Santander's association with Ferrari in Formula One racing, and BBVA's sponsorship of the Spanish football league and, making a strong bid for the American market, the NBA.

4   Enterprise Risk

Uncertainty is the key factor in any enterprise, and profit is the entrepreneur's reward or compensation for taking the risk that uncertainty poses.

Studies regarding the impact of risk on business Poch Torres [7] have proven two things: a) the market creates risk but does not manage it, and b) there is no such thing as zero risk.

Therefore, from an entrepreneurial perspective, any activity, including sponsorship, entails a degree of risk that may affect a company in different ways.

In functional terms, such activity exposes a company to financial risk, market risk, credit risk, legal risk, IT risk and organisational risk.

Sports sponsorship also has its own risks that need to be addressed, as any wrong decision can have disastrous consequences and the results may be the exact opposite of what was originally intended.

Each type of risk, as the complex phenomenon it is, presents a series of objective and subjective characteristics. Identifying, analysing and tracking these characteristics are part of the different phases of internal risk management within an integrated framework, which is essential for mitigating these risks and giving the company a greater competitive edge.

Sports sponsorship risk is a subtype in the category of corporate social responsibility (CSR) risks Meenaghan [8], yet its evaluation must consider much more than the impact of the company/brand in terms of reputation or consumer perception. It is usually considered a low-risk activity, but it has the potential to inflict major losses and damage the brand's image.

Failed sponsorships can result in huge losses and significantly tarnish a company's public image. Sports-related scandals are on the rise, and today there is a growing demand for clauses to guard against or mitigate potential losses in sponsorship contracts so that such scandals do not end up affecting the profit and loss accounts of the sponsor and/or sponsored parties.

The following list identifies some of the factors or situations that represent potential risks in sports sponsorship:

- Drug use, violence, fraud and other negative conduct (can tarnish the image of the sport and the brands associated with it)
- Overabundance of sponsors in a particular sport
- Need for the chosen sport to be consistent with the image the sponsor wants to project
- Some uncontrollable risks
- Inadequate planning
- Difficulty of quantifying the impact of sponsored initiatives and determining their true profitability
- The need for a major financial layout, as the sponsorship itself must also be publicised

Sports sponsorship also entails the risk of what we might call "war advertising", in which brands, fashion labels, football teams, automotive manufacturers and other players may become involuntary protagonists of armed conflicts reported by news agencies and television stations around the world. Companies that find themselves featured almost daily in war-related images must address this new reality and look for ways to minimise the damage and protect their image so that consumers do not end up associating the brand with violent war scenes, which can have an undesirable effect on its reputation.

5   Controlling, Managing and Evaluating Risks

Like all risks, sponsorship risk Harvey [4] must be managing, monitored and controlled in a constant, proactive process involving technical and human factors and internal control and oversight mechanisms. The primary purpose of this process is to create value for the company and give it an advantage over its competitors.

The generally accepted practice is to adopt a methodology based on the eight components which, according to the ISACA Framework, provide reasonable security and the ability to successfully weather or avoid crisis situations. These are: internal scoping, setting objectives, risk identification, risk evaluation, risk response, governance, information and communication, and supervision.

The risk management and control process must be a dynamic, well-structured programme that encompasses a number of open phases which can be applied to the general process as well as to a specific programme, activity or stage of the process. It should also be flexible enough to be useful at the strategic, tactical and operational levels, and its application will depend on the context in which it is used.

Once control measures to mitigate risks are in place, the next step is to re-evaluate the company's exposure to residual risk, which in some cases may still be high.

All of these risks need to be evaluated according to a matrix that takes all variables into account, such as security problems, choosing the wrong event and even outright failure, as well as any negative repercussions the event may have in political, cultural or environmental circles. A prescriptive risk evaluation method provides an unobstructed general view of the projects and establishes homogeneous criteria for evaluating risks, which are higher in some sectors than in others.

Proper evaluation of sponsorship is crucial for success, but this type of quantification has yet to become highly professionalised in Spain. We need better tools to guarantee the desired levels of audience, attendance and awareness, which can only be obtained using advanced techniques tailored to each particular situation, such as Spindex or SPR. If we fail to evaluate properly, we run the risk of investing in ineffective sponsorship and therefore wasting valuable resources.

The sport-related scandals that have come to light in recent years (Oscar Pistorius, Lance Armstrong, Tiger Woods and countless other cases of doping and corruption) prove that sports sponsorship has its risks, and that companies must have a strategy in place to ensure that the benefits (both financial and reputation-wise) outweigh the potential dangers of a sponsorship initiative.

6 Conclusions

The principal conclusion is that companies need to address the problems they will face when they choose to implement the new advertising and marketing tool of sports sponsorship.

Like any other enterprise, sports sponsorship creates risks that must be identified, analysed, evaluated, controlled, mitigated and managed in a business environment.

This effectively sums up our proposal: to emphasise the need for sponsorship risk management within the context of enterprise risks.

References

1. C. Campos, Sports Marketing and Sponsorship (Gestió i Promoció Editorial, Barcelona, 1997)
2. J. De Andrés, *Mecenazgo y Patrocinio: las claves del marketing del siglo XXI* (Editamex, Madrid, 1993)
3. M.A. Davara Rodriguez, *Estrategias de Comunicación en Marketing* (Editorial Dossat, Madrid, 1992)
4. B. Harvey, J. Advert. Res. **41**(1) (2001), pp. 93–123
5. J. Ibáñez, Investigación y Marketing **71** (2001), pp. 63–76
6. A. Mandado, Investigación y Marketing **83** (2004), pp. 60
7. R. Poch Torres, Gestió del control intern de riscos en l'empresa postmoderna: àmbits econòmic i jurídic. *Reial Academia de Doctors de Catalunya* (2010)
8. T. Meenaghan, Int. J. Advert. **10**(1) (1994), pp. 95–122

The Cultural Heritage Between Protection and Enhance

Alessandro Bianchi

Abstract My short speech is referring to one of the three topics on which the conference is focused: the culture. More precisely is referring to the tangible cultural heritage, ie the set of archaeological, architectural and artistic artifacts. As we know most of this heritage is located in Europe and in particular in Italy. We also know that it is now widely believed that this heritage is one of the greatest resources upon which we can base the development of our countries, including economic. Starting from these premises, I would like to bring to your attention three points.

Keywords Right to the culture • Culture and development • The case of Rome

1   First Point

The first one is a very general point and refers to the concept "Right to culture".

It may seem like a strange concept, but I invite you to reflect on the fact that today we usually speak of "Right to environment", and we recognize it as the right of citizens to live in a sustainable environment, where air and water are not polluted, the landscape is protected, the waste is disposed of properly, mobility is sustainable and so on.

Laws, that affect those who do not respect them, now guarantee this right but this is true only for 40 years ago.

Maybe someone of you remembers the sinking of the oil tanker Torrey Canyon in the North Sea, which took place in 1971.

A. Bianchi (✉)
Università Mediterranea, Reggio Calabria, Italy
e-mail: ale.bianchi7@fastwebnet.it

A.M. Gil-Lafuente and C. Zopounidis (eds.), *Decision Making and Knowledge Decision Support Systems*, Lecture Notes in Economics and Mathematical Systems 675, DOI 10.1007/978-3-319-03907-7_4, © Springer International Publishing Switzerland 2015

From that first environmental catastrophe sprang a long process that led to identify and defend the right of citizens to the environment, in the sense that I said before.

And so I believe that there is now a mature awareness of the extraordinary importance that cultural heritage plays in the world, and therefore I think there are the conditions to make pervasive the concept of the "Right to culture", and to define a Charter of this right.

In fact this Charter is already contained in the Declaration of Fribourg (CH) signed by many institutions and personalities in May 2007.

The Article 1 of this Declaration says:

Cultural rights are essential to human dignity and they are an integral part of human rights; to these rights should apply the principles of universality, indivisibility and interdependence.

In my opinion it's necessary that from this statement starts a process that leads to define the concept of the "Right to culture" and to introduce rules that guarantee its exercise.

2   Second Point

The second point I want to make, deals with the culture as an engine of development.

I believe that this concept is now accepted by all, because there are many examples of the fact that the enhancement of cultural heritage has increased the production and trade of goods, has set in motion synergies with scientific research, higher education and technological innovation, has created wealth and new employment.

I could cite at least three examples well known to you: Barcelona, Bilbao and Valencia. But there are also the examples of Glasgow, St. Petersburg Dublin and many more.

In all these Cities have been implemented development strategies based on the enhancement of cultural heritage, and the results were excellent.

However there is still a limiting factor for the complete unfolding of this strategy. This factor is present in a particular way in Italy: it is the contrast between two different points of view.

The first point of view is that of those who believe that cultural heritage only needs to be protected, because the enhancement can lead to its degradation.

This is the ruling opinion of the Superintendents.

The second one is that of those who propose enhancement measures, without attention to the problem of protecting.

This is the ruling point of view of the Builders.

As always we need to find a balance, because there is no doubt that the need to protect the cultural heritage constitutes the first necessity.

But we must be aware that if the cultural heritage is not made accessible to citizens and is not put to value, it is not clear in favor of those being protected.

This is a problem that I wanted to mention because it's strongly felt in Italy and often leads to a deadlock of interventions.

We must learn to overcome this block, because it is contrary both to the needs of the cultural heritage and the interests of citizens.

3   Third Point

The third point is not a subject for debate, but the presentation of a study case, which in my opinion is extremely significant with regard to the protection and enhancement of cultural heritage.

It's the case of Parco dei Fori in Rome.

Currently these areas and these precious monuments are separated from one another, suffer the siege of traffic and are exposed to degradation by pollution.

But the worst thing is that it has been broken the continuity between Foro Romano and Fori Imperiali.

In recent months, during the election of the Mayor of Rome, the Progetto Roma Association has launched two proposals:

1. Eliminate the road built in 1931–1932 and restore the unity of Fori.
2. Start an international ideas competition for the creation of the Parco dei Fori, ie the largest archaeological park in the world.

This is also my opinion, and I think that Parco dei Fori in Rome is one of the most important examples to which we must to give our attention when we talk about the role of cultural heritage in twenty-first century.

Uncertain Optimal Inventory as a Strategy for Enterprise Global Positioning

González S. Federico, Flores R. Beatriz, Anna Maria Gil-Lafuente, and Flores Juan

Abstract This paper solves the classical and fuzzy Economic Order Quantity (EOQ) problems. We consider all parameters to be uncertain, expressed as triangular fuzzy numbers. We present a comparative analysis between both methodologies, presenting competitive advantages of one versus the other one. We also present how these strategies can be applied to a company in search of a better position in the world market, aiming to make the world-class enterprises.

Keywords Inventory • Uncertainty • Fuzzy logic • Business improvement

1 Introduction

Kaufmann A. and Gil Aluja J. [1], state that the production process in a company has been considered as the center around which the company's activity spins. The production process requires raw materials and finished products supplies at critical times. This fact requires the design of an efficient and effective program to deliver raw materials to the productive process; otherwise, the plant could become inactive for the lack of them. The lack of raw materials implies to support high costs for not operating at capacity levels established to meet demand.

G.S. Federico (✉) • F.R. Beatriz
Facultad de Contaduría y Ciencias Administrativas, Universidad Michoacana de San Nicolás de Hidalgo, Edificio A-II. C.U. Morelia México, Mexico
e-mail: fsantoyo@umich.mx; betyf@umich.mx

A.M. Gil-Lafuente
Faculty of Economics and Business, University of Barcelona, Barcelona, Spain
e-mail: amgil@ub.edu

F. Juan
Facultad de Ingenieria Electrica, Universidad Michoacana de San Nicolás de Hidalgo, Edificio Omega. C.U., Morelia México, Mexico
e-mail: juanf@umich.mx

A.M. Gil-Lafuente and C. Zopounidis (eds.), *Decision Making and Knowledge Decision Support Systems*, Lecture Notes in Economics and Mathematical Systems 675, DOI 10.1007/978-3-319-03907-7_5, © Springer International Publishing Switzerland 2015

Entrepreneurs keep stocks (inventories) of raw materials, supplies and finished goods when they could place their assets in other productive activities, thus immobilizing money for the following reasons listed:

- The existence of productive activity makes inevitable to maintain specified inventory levels.
- Given that future is considered uncertain, in many cases it is not possible to accurately forecast demand, so it is necessary to have an optimal minimum inventory level that meets unexpected fluctuations in demand, looking for a cost which tends to zero for this activity.
- Speculations arise when a sudden price increase is expected or there is a high possibility that sales will increase in the future, so you have the possibility of high returns on investments.

According to Narasimhan S. et al. [2], inventory control is a critical aspect of successful administration. When maintaining inventories implies a high cost, companies cannot afford to have excessive amounts of money invested in inventory.

To minimize product inventory, the company requires a flawless strategic plan seeking to match supply and demand levels, such that the stocks of products are minimal.

According to Schroeder R. [3], the inventory is regarded as commodity stocks kept in one place, at one time. The inventory is a stored amount of materials used for production and to meet consumer demands.

The basic decisions that have to be faced in inventory management are, among others:

- When to place orders.
- How much to order.

To answer those questions it is necessary to know the behavior of the expected demand for the period of time in which you want to perform the analysis. The number of days to consider in the analysis depends on the annual cost to maintain the inventory (h), commonly fixed taking into account a percentage of the cost of the product (item), item cost (C), and the enabling costs (S).

Hence, inventory management can be considered as one of the most important administrative functions of production, since it requires an important amount of assets. An improperly managed inventory affects the delivery of products to consumers. The optimal management of inventories in the company has a strong impact, particularly in the production, marketing, and finance.

In this sense, the operational guidelines for an appropriate inventory management are:

- Financial management seeks to maintain inventories at a low level to avoid excess inventory.
- Marketing aims to have high levels of inventory to ensure sales.
- Operation seeks to have proper inventory management to ensure efficient production and homogeneous employment levels.

The company must strengthen the management of inventory systems to balance the previous guide lines that seem to be in conflict, so it is necessary to determine an optimal inventory size to meet the market needs.

In an inventory system there is uncertainty in the behavior of the supply and demand, and the time involved in the process up to the consumption stage. Managers maintain inventory levels in the company to deal with and protect against these uncertainties and maintain good operation of the company in the market.

So the problem to be addressed in this work is to determine how much to order, to deal with uncertain demand, as well as to determine the time to order. To this end, we need to consider:

2 Inventory Costs

In the inventory cost structure we have the following types of costs: product cost, ordering cost, inventory cost, and the stock-outs cost.

1. Item Cost. This is the cost of buying or producing individual items. It is generally expressed as a unit cost multiplied by the stock capacity.
2. Cost of Order, losses or preparation. This cost is related to the acquisition of a group or batch of items. The ordering cost does not dependent on the number of produced items; it is assigned to the whole batch.
3. Inventory Costs. These costs are associated with the permanence of items in stock during a period. This cost is usually charged as a percentage of the item value per unit of time.

Inventory costs have three components commonly expressed as:

- Equity Cost. When items are in stock, the invested asset is not available for other purposes. These represent a loss of opportunities for other investments; the inventory cost is therefore an opportunity cost.
- Storage Cost. This cost includes variable space, insurance, and taxes.
- Cost of obsolescence, deterioration, and loss. These costs are assigned to products that have a high risk of becoming obsolete; the greater the risk, the higher the cost. Perishable products are assigned the costs of rotting. For food, loss costs include costs related to theft and shelf damage.
- Stock-outs Costs. This factor reflects the costs incurred when the company runs out of items in stock. The absence of raw materials, supplies, and finished product, when they are required, result in a loss of profitability and business opportunity in that instant or future business associated caused by the customer wait. This loss of opportunity is assessed as a cost of absence.

3   Demand Behaviour

According to Kaufmann A. and Gil Aluja J. [1], Gonzalez S. F. and Flores R. B. [4], future demand behavior may exhibit the following characteristics:

- When the company knows exactly how it will behave in time; this is deterministic behavior, given under certainty.
- When the company does not know how it will behave in time; this is not deterministic, probabilistic, or stochastic behavior.
- When the company does not know what levels it will reach, although it is not completely ignorant of it. It is therefore possible to use the experience in this environment of uncertainty.

4   Classical Economic Order Quantity (EOQ)

This methodology was developed by F. W. Harris in 1915. Today it is still widely used in business practice for inventory management, when demand is independent.
The basic assumptions used in developing the model are:

- The demand level is known, and constant over time.
- The delivery time for items ordered at a given instant is zero; it is known and constant (with a fixed number of days).
- No stocking is allowed, since demand and lead time are constant. It is possible to determine exactly the time to make a purchase of raw materials and supplies to avoid stock-outs.
- Raw material is acquired or produced in groups or batches, and batches are placed in stock all at once.
- It uses a cost structure as follows: the unit cost of the item is constant and there are no discounts for large purchases.
- The fixed cost for ordering items is (k) monetary units.
- The item unit cost is (c).
- The item (product) in analysis is unique, there is no interaction with other products.

Given these assumptions, the inventory level behavior for the case shown in Fig. 1:
Let us consider:

Q = Ordering Quantity (No. of units).
d = Demand (No. of units/time).
K = Fixed cost.
c = Unit cost ($/item).
h = Cost of holding one unit in stock ($/article) = i %(c).

Figure 2 shows graphically the cost behavior.

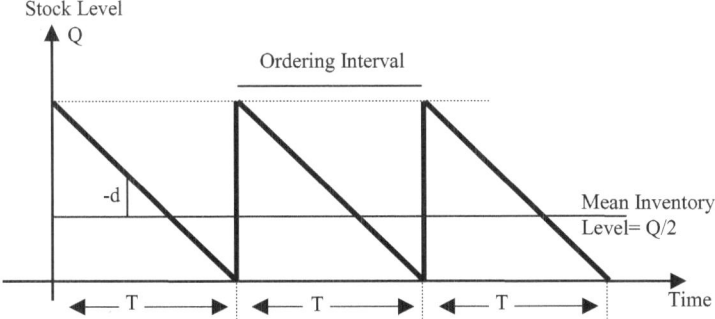

Fig. 1 Classical EOQ

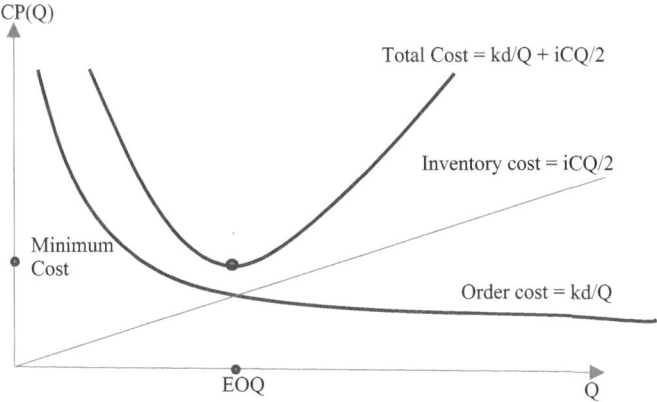

Fig. 2 Cost behavior

Figure 2 shows that as Q increases, the purchase cost decreases, due to fewer orders placed annually. At the same time, the inventory cost component increases because inventory average increases. This causes the costs of buying and holding stock to compensate; one decreases while the other one increases.

To determine the value of Q that minimizes the cost CP (Q), the partial derivative of CP (Q) is computed and equaled to zero, solving for Q, as shown:

The cost per period is:

$$CT(Q) = (\text{Fixed Cost}) + (\text{Materials Cost}) + (\text{Stock Cost}). \quad (1)$$

Using the above terms, this cost can be expressed as:

$$CT(Q) = k + cQ + h\left(\frac{Q}{2}\right)T. \quad (2)$$

Where the Optimal Mean Cost can be expressed as:

$$CP(Q) = \lim_{n \to \infty} \left[\frac{nCT(Q)}{nT} \right] = \frac{CT(Q)}{T} = \frac{k + cQ + h\left(\frac{Q}{2}\right)T}{T} = \frac{kd}{Q} + cd + \frac{hQ}{2}. \quad (3)$$

Where $T = Q/d$.

Optimal Mean Cost $CP(Q^*)$, can be obtained computing its partial derivative with respect to Q. The place where this derivative is zero:

$$\frac{\partial CP(Q)}{\partial q} = 0 = -\frac{kd}{Q^2} + \frac{h}{2} = 0. \quad (4)$$

Therefore:

$$\frac{kd}{Q^2} = \frac{h}{2}$$

$$Q^2 = \frac{2kd}{h} \quad (5)$$

$$Q^2 = \sqrt{\frac{2kd}{h}}.$$

The above equation for (Q) represents the order size that minimizes the average cost of the inventory operation. To calculate (Q), we use the annual basis, nevertheless, we can use any time unit as long as demand rates and interest are compatible.

To determine the time required to make inventory zero, we use the following equation.

$$T = \frac{Q}{d} = \sqrt{\frac{2k}{hd}}. \quad (6)$$

To determine the Optimal Stock Mean Cost, we use:

$$CP(Q) = \frac{kd}{Q} + cd + \left(\frac{hQ}{2}\right). \quad (7)$$

5 Analysis Case

As an example, consider the company "Procuramiento Industrial" (procurement or supply) of the San Rafael Hospital of Morelia, Michoacan, Mexico. The general manager of the company must determine how and when to order to ensure availability of X-ray film (in the radiology department) to provide high quality service, and not to provide delayed services to patients.

The aim is that the hospital never run out of this product, which is considered strategic (critical), while keeping the total cost as low as possible.

Consider that the X-ray film provider is located outside the city of Morelia, items are replaced in batch orders at a rate of 150 films/month (according to the hospital's operating history), it has a deterministic behavior, the film delivery time is 1 week and deficit is not accepted.

The hospital's accounting department has provided the following information:

- The fixed reorder cost is $100 currency units (pu).
- The purchase cost (no discount for large amounts) of each film is $20 monetary units (pu).
- It has a transfer rate of 30 % per year ($i = 0.3$) to reflect the cost of storing the film in a special area, as well as the opportunity cost of money tied up in idle inventory.

The manager is interested on determining the Optimal Order Quantity (Q) and the new orders period.

To determine the order quantity, standardize the information; considering 52 weeks/year:

- Annual demand $d = (1{,}500$ films/month$)\, 12 = 18{,}000$ films/year.
- Time: $t = 1$ week $= 1/52$ years.
- Annual transfer rate, $i = 0.3$.
- Fixed Cost Order $k = \$100$ per order.
- Order Cost $c = \$20$ per film.
- Annual Conservation Cost: $h = i\,(c) = 0.3\,(20) = \$6$ per film per year.

Therefore, the order quantity is:

$$Q^2 = \sqrt{\frac{2kd}{h}} = \sqrt{\frac{2(100)18000}{6}} = 774.6 \cong 774 \text{ o } 775 X - \text{ray films.} \qquad (8)$$

For $Q = 774$, the Annual Cost is:

$$CD(Q) = \frac{kd}{Q} + cd + \left(\frac{hQ}{2}\right). \qquad (9)$$

$$CD(Q) = \frac{100(18000)}{774} + 20(18000) + \left(\frac{0.3(20)774}{2}\right) = 364647.58.$$

For $Q = 775$, the Annual Cost is:

$$CD(Q) = \frac{kd}{Q} + cd + \left(\frac{hQ}{2}\right). \qquad (10)$$

$$CD(Q) = \frac{100(18000)}{775} + 20(18000) + \left(\frac{0.3(20)775}{2}\right) = 364647.58.$$

For general decisions the manager, always opts for choosing the order quantity that provides the minimum total cost. For the cases of 774 or 775 films, the total cost is the same: \$364,647.58. The manager decides order 775 films.

The average number of orders/year is:

$$NP = \frac{d}{Q} = \frac{18000}{775} \cong 23.23. \tag{11}$$

The time between orders is:

$$T = \frac{Q}{d} = \frac{775}{18000} \cong 0.043. \tag{12}$$

To calculate the reorder point for new orders the optimal order quantity is $Q = 775$ in average, 23 times a year. We need to place orders so that the inventory level is reaching zero when the new period arrives. The lead time for orders is 1/52 years.

How many movies are needed in stock to meet the demand during lead time while the order is placed?

How: $d = 18,000$ movies/year.

Therefore, the number of films needed during the lead time is (given that $d = 18,000$ films/year).

$$R = \frac{d(\text{lead time in days})}{365 \text{ days}} = \text{units.} \tag{13}$$

The Reorder Point is:

$$R = d(t) = 18000\left(\frac{1}{52}\right) = 346.15 \cong 346 \text{ films.} \tag{14}$$

This implies that when the inventory reaches 346, we will place the new order of 775 films.

6 Fuzzy Economic Order Quality (EOQ)

According to Guiffrida A. L. [5], the model of the EOQ in an environment of uncertainty, using Fuzzy Logic, considering (F) as shown in Table 1.

Table 1 EOQ analysis

Model	Model parameters			
	Q	c	h	d
Park (1987)	*	F	F	*
Vujosevic (1996)	*	F	F	*
Lee and Yao (1999)	F	*	*	*
Yao et al. (2000)	F	*	*	F
Yao and Chiang (2003)	*	*	F	F
Wang et al. (2007)	*	F	F	*

* is = No fuzzy analysis includes
Source: Alfred L. Guiffrida, Kent State University, Kent, Ohio. Chap 8. Fuzzy Inventory Models (2010)

In business life analysts and consultants generally rely on all those techniques that encompass everyday phenomena in all its complexity in order to formalize them and act on them [6–8].

This perspective has traditionally lead reasoning based on the concept of precision and frequently, were performed quantitatively through classical schemes of mathematics. This resulted in a modification of reality to adapt to the mathematical models, rather than the reverse (i.e., an adaptation of the models to the real facts).

Thus reality, though accurate in itself, could only be captured through some of its aspects, which has led to simplification, and precision has been removed. Such findings are forcing us to develop models based a multivalent logic environment using Fuzzy Logic. These models provide a very good approximation to the exhibited behavior of the studied phenomenon.

In the analysis we will consider all parameters to behave uncertainly (fuzzy), derived that under real circumstances, companies dwell in uncertain environments, and all the elements that have to do with accounting and financial indicators are fuzzy. Even more, time is considered a nonrenewable resource. Therefore, all elements involved in the analysis are expressed as triangular fuzzy numbers. The parameters are expressed as:

$$\tilde{d} = (17000, 18000, 19000). \tag{15}$$

$$\text{Time} = \tilde{t} = \left(\frac{1}{50}, \frac{1}{52}, \frac{1}{53}\right) \text{years}.$$

$$\tilde{i}\% = (0.2, 0.3, 0.4)$$
$$\tilde{k} = (80, 100, 120)$$
$$\tilde{c} = (18, 20, 22)$$
$$\tilde{h} = \tilde{i}\%(\tilde{c})$$

To convert the classical EOQ model to a fuzzy one, we do the following. To determine the time required to take the stock to zero, we make use of the following equation.

$$\widetilde{Q} = \sqrt{\frac{2kd}{\widetilde{h}}}. \tag{16}$$

The Inventory Optimal Mean Cost is:

$$\frac{\widetilde{CP(Q)} = \widetilde{kd}}{\widetilde{Q} + \widetilde{cd} + \left(\widetilde{\frac{hQ}{2}}\right)} \tag{17}$$

The Mean Number of Orders/year is:

$$\frac{\widetilde{NP} = \widetilde{d}}{\widetilde{Q}} \tag{18}$$

Time between orders:

$$\frac{\widetilde{T} = \widetilde{Q}}{\widetilde{d}.} \tag{19}$$

The analysis makes use of a 11-value scale and the membership function of fuzzy numbers range from 0 to 1. So that to each level $[0 \le \alpha_k \le 1]$ corresponds a confidence interval $[0 \le \alpha_k \le 1]$ which can be expressed in terms of α_k, as follows:

$$\left[r_k^\alpha, s_k^\alpha\right] = [r + (m - r)\alpha_k, s - (s - m)\alpha_k]. \tag{20}$$

Therefore confidence intervals (k, i%, c, d) are expressed as:

α	$\widetilde{k}$	
0.0	80	120
0.1	82	118
0.2	84	116
0.3	86	114
0.4	88	112
0.5	90	110
0.6	92	108
0.7	94	106
0.8	96	104
0.9	98	102
1.0	100	100

α	$\widetilde{i\%}$	
0.0	0.20	0.40
0.1	0.21	0.39

(continued)

0.2	0.22	0.38
0.3	0.23	0.37
0.4	0.24	0.36
0.5	0.25	0.35
0.6	0.26	0.34
0.7	0.27	0.33
0.8	0.28	0.32
0.9	0.29	0.31
1.0	0.30	0.30

α	$\widetilde{c}$	
0.0	18.0	22.0
0.1	18.2	21.8
0.2	18.4	21.6
0.3	18.6	21.4
0.4	18.8	21.2
0.5	19.0	21.0
0.6	19.2	20.8
0.7	19.4	20.6
0.8	19.6	20.4
0.9	19.8	20.2
1.0	20.0	20.0

α	$\widetilde{d}$	
0.0	17,000	19,000
0.1	17,100	18,900
0.2	17,200	18,800
0.3	17,300	18,700
0.4	17,400	18,600
0.5	17,500	18,500
0.6	17,600	18,400
0.7	17,700	18,300
0.8	17,800	18,200
0.9	17,900	18,100
1.0	18,000	18,000

The order size, for $[0 \leq \propto_k \leq 1]$ is:

$\widetilde{Q} = \sqrt{\frac{2k\widetilde{d}}{h}}$	
555.959	1,125.463
574.327	1,080.293
593.335	1,038.013
613.027	998.316
633.449	960.939
654.654	925.658
676.697	892.275

(continued)

699.641	860.619
723.553	830.540
748.511	801.905
774.597	774.597

$$\widetilde{Q} = (555.959, 774.597, 1,125.463).$$

For $[0 \leq \propto_k \leq 1]$, the Mean Cost is:

$\dfrac{\widetilde{CP(Q)} = \widetilde{kd}}{\widetilde{Q} + \widetilde{cd} + \left(\dfrac{\widetilde{hQ}}{2}\right)}$	
308,209.12	427,053.06
313,615.52	420,495.48
319,072.80	414,015.50
324,581.57	407,609.83
330,142.50	401,275.61
335,756.30	395,010.31
341,423.72	388,811.70
347,145.62	382,677.81
352,922.89	376,606.86
358,756.52	370,597.26
364,647.58	364,647.58

$$\widetilde{CP(Q)} = (308,209.12, 364,647.58, 427,053.06).$$

The Mean Number of Orders per year is:

$\dfrac{\widetilde{NP} = \widetilde{d}}{\widetilde{Q}}$	
15.10	34.18
15.83	32.91
16.57	31.69
17.33	30.50
18.11	29.36
18.91	28.26
19.72	27.19
20.57	26.16
21.43	25.15
22.32	24.18
23.24	23.24

$$\widetilde{NP} = (15.10, 23.24, 34.18).$$

The time between orders is:

$\widetilde{T} = \dfrac{\widetilde{Q}}{\widetilde{d}}$	
0.029	0.066
0.030	0.063
0.032	0.060
0.033	0.058
0.034	0.055
0.035	0.053
0.037	0.051
0.038	0.049
0.040	0.047
0.041	0.045
0.043	0.043

$$\widetilde{T} = (0.029, 0.043, 0.066).$$

7 Results

This section presents the results using a comparative approach, contrasting the classical the fuzzy EOQ models. The results are shown in Table 2.

The results reflect that using fuzzy logic analysis, one obtains a broader spectrum of the results in comparison to the classical theory, which hides information. We therefore allow the entrepreneur to take better decisions for a more efficient and effective operational planning, which will take him/her to position the organization as a world-class company.

It is important to note that in the case presented in this paper, the result obtained using classical methodology for EOQ, matches the Fuzzy EOQ. However, the fuzzy analysis provides a wider range of results, making highly possible to include the real operational dynamics of the company.

8 Conclusions and Recommendations

From the above we can conclude that the application of Fuzzy Logic presents a competitive advantage in this kind of analysis, given that the type of information found in this type of problems exhibits a dynamic behavior. Therefore, the application of Fuzzy methods for treating such problems in the company is recommended. This methodology will enable the company to gain a competitive advantage in relation to companies using the classical methods.

Table 2 Classical vs. fuzzy EOQ

Variable	Classical	Variable	Fuzzy
Q	774	$\widetilde{Q}$	(555.959, 774.597, 1, 125.463)
$CP(Q)$	\$364,647.58	$\widetilde{CP(Q)}$	(308,209.12, 364,647.58, 427,053.069)
NP	23.23	$\widetilde{NP}$	(15.10, 23.24, 34.18)
T	0.043	$\widetilde{T}$	(0.029, 0.043, 0.066)

References

1. A. Kaufmann, J. Gil Aluja, *Introducción de la teoría de subconjuntos borrosos a la gestión de las empresas* (Velograf S.A, España, 1986)
2. S. Narasimhan, D.W. Mc. Leavey, P. Billington, *Planeación de la producción y control de inventarios* (Prentice Hall, México, 1996)
3. R.G. Schoeder, *Administración de operaciones. Toma de decisiones en la función de operaciones* (McGraw Hill, México, 1992)
4. F. González Santoyo, B. Flores Romero, *Teoría de Inventarios en la empresa (notas de seminario)*. Doctorado en Economía y Empresa, Universitat Rovira i Virgili, España, 2002
5. A. Guiffrida, in *Inventory Management: Non-Classical Views. Chapter 8*, ed. by M.Y. Jaber (CRC, Boca Raton, FL, 2010), pp. 173–190
6. F. González Santoyo, B. Flores Romero, A.M. Gil Lafuente, *FeGoSa-Ingeniería Administrativa S.A. de C.V.* (UMSNH, IAIDRES, Morelia, México, 2010)
7. F. González Santoyo, B. Flores Romero, A.M. Gil Lafuente, *Procesos para la toma de decisiones en un entorno globalizado* (Editorial Universitaria Ramón Areces, España, 2011)
8. A. Kaufmann, J. Gil Aluja, G.A. Terceño, *Matemáticas para la economía y la gestión de empresas* (Foro Científico, Barcelona/España, 1994)

Robotics, Automation and Information Systems: Future Perspectives and Correlation with Culture, Sport and Life Science

Yuriy P. Kondratenko

Abstract This paper devoted to development of most prosperous scientific directions in twenty-first century. Special attention paid to analysis of future perspectives of robotics, automation and information science and their correlation with life science and most important fields of human activity: sport and culture. Different kinds of sport robots and their functionalities are discussed in details, in particular, humanoids, home service robots, game robots etc. The literature analysis shows: some research and investigations deal with problem to find culture differences when the user interacts with a home robot having facial expressions; different nations have different cultures which are inevitably reflected in communication modes; in the field of sports and athletics, it is important for athletes to recognize their own performance by themselves to gain skills effectively using modern robotic and information systems. The role of information technology in culture and sport development is also under consideration as Internet, World Wide Web, and related information and communication technologies have rapidly spread to a large number of countries and cultures.

Keywords Robotics • Automation • Information systems • Culture • Sport • Life Science • Correlation • Perspectives

Your Excellency,
   President of the Royal Academy of Economic and Financial Sciences (RACEF) of Spain Dr. Jaime Gil Aluja and RACEF's Members,
   President of Royal Academy of Doctors Dr. Alfredo Rocafort Nicolau,

Y.P. Kondratenko (✉)
Royal Academy of Doctors, Barcelona, Spain

Regional Inter-University Centre, Mykolaiv, Ukraine

Intelligent Information Systems Department, Petro Mohyla Black Sea State University, 68-th Desantnykiv Str., 10, 54003 Mykolaiv, Ukraine
e-mail: y_kondrat2002@yahoo.com

A.M. Gil-Lafuente and C. Zopounidis (eds.), *Decision Making and Knowledge Decision Support Systems*, Lecture Notes in Economics and Mathematical Systems 675, DOI 10.1007/978-3-319-03907-7_6, © Springer International Publishing Switzerland 2015

Distinguished Guests, Ladies and Gentlemen,

It is great pleasure for me that I have possibility to wish all of you great success on the occasion of this VIII International Act "Science, Culture and Sport in 21st century".

Let me cordially deliver a great friendship's feeling from the professors and students of Petro Mohyla Black Sea State University of Ukraine, from Regional Inter-University Centre in Mykolaiv and from Association of Ukrainian universities.

As representative of the Ukraine I think that today according to total globalisation and internationalization in different areas of human activities a lot of the same challenges and perspectives are common for different countries in the development of Science, Culture and Sports.

First of all it concern creating modern partner relations in main priority directions of science, exchanging results and experience, taking into account cultural peculiarities of different people, international approach and national specific in sport industry.

I believe that all perspective directions in science, culture and sport will achieve new and more high level of their development in twenty-first century.

1 Correlation Between Science, Culture and Sports

The topic of our VIII International act is very important for sustainable development of each society and world values in whole.

- If we speak about "science" we should take in to account different fields of science, especially, Engineering, Economics, Medicine, Pedagogy, Agriculture and so on.
- If we speak about "culture" we can consider Literature, Music, National traditions, Languages and so on:
- If we speak about "sport" we should take into account different kind of sport competitions in football, tennis, snooker and so on.

As example, Nanotechnology and Nanomaterials will be developed very intensively in twenty-first century [1]. Some prognosis show that this direction has very important perspectives in different fields of life science, especially in medicine for early diagnostics.

The contribution of Nanotechnology (NT) and Nanomaterials (NM) for sports and physical culture is very significant because NT and NM has been applied in sports widely (human movement science, sports halls, facility, equipment etc.).

But at the same time NT and NM maybe cause some potential adverse effects on body and environment. The special biological effect and the safety should not be neglected.

Human being should increase the basic research and epidemiological investigations about environment-health-safety of nanomaterials, to provide accurate exposure levels and the guiding principles of nanoparticles risk, to provide the necessary theoretical basis for establishing the standard of nanomaterials production and work place environment.

With the attention of the whole world scientific potential, no doubts, that the safety problems for NT and NM will be solved finally in twenty-first century.

In this presentation I'd like to consider the perspectives of development and research of such engineering domains as Robotics, Automation and Information Technologies according to their intersection and correlation with sports and culture in more details.

2 Perspectives of Robotics and Its Interaction with Culture and Sport

There are many approaches to sport from different sciences and engineering.

Robotics is a relatively new area and has had moderate attention from sport specialists.

Humanoids are the most advanced robots [2, 3] and have been expected to act in various fields including education.

It is important to investigate the meaning of the word "humanoid robots" in different cultures from the point of view psychological and engineering perspectives. Students from Japan, USA and Korea gave as experts different images and facilities of humanoid robots.

467 participants from 7 different countries (Japan, Mexico, USA etc.) also have different opinions which was published as answers for 14 questions in three clusters [3]:

- Attitude towards interaction of robots.
- Attitude towards social influence of robots.
- Attitude towards emotions in interaction with robots.

So, the participant's cultural background had a significant influence on their attitudes.

According to opinion of Luc Steels [4], director of the Sony Computer Science Laboratory Paris, the artificial intelligence should work like scientific field where can be progressive accumulation of knowledge which can be used in future for creating robot culture.

In the field of sports and athletics, it is important for athletes to recognize their own performance by themselves to gain skills effectively.

For example, it is difficult for non-professional swimmers to understand how they swim.

The underwater robotic vehicles and underwater applications, the methodology of presenting information from robots to humans in underwater environments has become an important science direction.

The buddy robot as swim support system called Swimoid [5] can swim directly under the user and present information through the display mounted over the main body.

Sport robots designing based on biomechanics [6] have very strong perspective. Among them different type of sport robots:

- Serving balls.
- Helping to provide sport training.
- Substituting humans during training.
- Physically participating in competitions.
- Physically participating in competitions against humans.
- Serving as models of real sport performance.
- Helping organizers of sport events.
- Robot toys.

The development and research of NXT-robots [7] for education in the field of computer programming, for example, to teach Java-based concurrency, as well as game robots [8–11] have a lot of perspectives in twenty-first century, specially:

- Robotic table tennis based on fuzzy decision method.
- Semi-autonomous competitive robots for games of sumo-wrestling and soccer.
- Wireless communication in robot soccer.
- Statistical analysis from measurement of eye movements for tactics of air hockey robot.
- A water walking robot inspired by water strider.
- Billiard robot in pulling game training.
- And others.

Interaction between human and the home service robot on a daily life cycle is one among important research directions taking into account the necessary to explore the perceived roles of home service robots between different cultures.

From the publishing works [12–14] it is known that human preferences on interaction modes and features of the home service robots are different in different countries and results of the study in Japan and Taiwan indicated a complicated issue related to the acceptance and reliability of the social robots.

In particular, participants from Taiwan and Japan reflected their preferences on communication matters in function, service, interaction, appearance, likability and trust issues. Participants from both cultures preferred home service robot to provide information over social services, while Taiwanese participants possessed significant preference on scheduling (planning) function.

For the future needs of home service robots Taiwanese participants preferred entertainment service robots, and Japanese participants preferred caretaker robots.

Communications between humans and robots is a very critical step for the integration of social robots into society. Emotion expressing through a robotic face is one of the key points in communication.

Some research and investigations deal with problem to find culture differences when the user interacts with a home robot having facial expressions.

Different nations have different cultures which are inevitably reflected in communication modes.

Without knowledge of these differences in crosscultural communications, people may find it hard to pass on the desired message and misunderstandings may arise.

In order to let the home robot supply various services with no national boundaries it is necessary to consider different cultures during conversation. Some investigations [15, 16] of differences in Chinese, Japanese, US and Western cultures when the user communicates with home robot show the importance of such research direction.

So, home robots in future have interact in a social environment, with people with different cultural background.

Development of robotic systems for various clinical problems currently takes on special significance in the field of clinical device-making industry and development of technical gear for public health service [17].

The analysis of verbal and non-verbal events in robotic surgery in two different surgical teams (one in the US and one in France [18]) revealed differences in workflow, timeline, roles, and communication patterns as a function of experience and workplace culture.

There is an increasing trend in using robots for specific medical purposes, specially for the rehabilitation of sportsmen [19, 20]. For instance, to arrange doing exercises for patients (sportsmen) with diseases of the support and locomotion system (SLS), the considerable number of medical staff is necessary.

The considered in [20] robotic system ensures ability to fulfill exercises for working up moving function of patient's extremities in self-training regime.

One of the peculiarities of the considered system is its sensing by the method of myoelectric control [21] for monitoring of trained SLS behavior and correction of actuators' parameters of robotic systems.

Myoelectric signal, which are sensing the robotic systems, is extended from skeleton muscle engaged in exercises. Earlier the myoelectric principle was used for developing the prosthetic device for upper extremities and distance master-slave manipulator.

Such kind of robotic system for rehabilitation tasks is presented in [20, 21].

This system consists two robots—R1 and R2, connected with a control unit (CU) [21].

The output signals of position sensors (PS) of both robots' actuators, as well as output signal of myoelectric correction unit (MCU), are passed to CU inputs.

Outputs of CU are connected with tracking drives (TD) of each degree of freedom of both R1 and R2 robots.

To arrange visual feedback the robotic system (RS) is equipped with video monitor (VM).

Gr1 and Gr2 grasps of robots are fixed at adjoining joints of trained extremities, at the corresponding skeleton muscle the surface electrodes are set for monitoring of myoelectric signals.

The initial position of trained extremity and various target positions to move are reflected at the monitor.

Target positions are distinguished of each other by the implementing exercises' hardship level and, as a rule, the i-th zone is changed by (i + 1)-th only after analysis of myoelectric signals and character of its changes, which corresponds to tiredness of trained muscles.

CU consists of two control levels: tactical and executive. Interaction between CU and MCU is implemented at executive level.

The tactical control level fulfills behavior planning of executive gear of R1 and R2 robots depending on initial and gain position of trained extremity.

The gesture communication [22] between human and social robot is also very perspective approach in the case of users with different culture.

Problematic of machine vision is so attractive to modern scholars, that the hardware capabilities, developed using the latest advances of electronics and computer engineering, has reached the level which is close to human "specifications".

And so, in the way to realization the fantastic dream of future, namely the creation of complicated free-running robotechnical complexes—"intelligent machines" that function interactively, we need to deal with only one fundamental problem—developing methods and algorithms for images "understanding" [17, 23].

There are only a few systems that perform functions of man-machine interface by using gestures on the service market today.

For example, a device management system created by Mgestyk Technologies, which operates with the help of a special 3D camera and is used mostly for computer games. The main disadvantage of this system is the need of expensive specialized cameras.

Under pressure of the circumstances, the actual task of scientific development is recognition of gestures using ordinary web-cameras, which are widely used and have a reasonable price.

Using the term "recognition" we mean not only the selection gesture, but also its identification for future using (data transfer or command execution).

There are some commercial exercise machines used for rehabilitation purposes [19–21]. However, their controlling systems are designed for healthy people.

In such systems the primary user of the system is operator (doctor, physiotherapist), who completely controls it.

The patient represents as the final link in the system.

Increasing productivity of rehabilitation system could be reach by improving human-machine interaction.

Such system should consist video-gesture recognition subsystem, which requires minimal effort from the user (patient).

Thus, the patient himself can manage to operate the basic functions of the robot (change the speed, intensity of exercise or stop it).

It is necessary to "show" how does the command work to system, before the beginning of exercises, commands, to establish a correspondence between gestures and commands or put it simply: "what gestures for which command will meet."

General algorithm of gestures management system includes following steps [22]:

1. Image capture is made by using web-camera, which is connected to the personal computer (PC) and transfers sequence of frames in real time to the program.
2. Detecting the hand on the image—on the image, received by web-camera using, hand region is detected. Detection, usually, consists of two stages: segmentation (selection) of color, which is used for reducing the amount of processing information, and areas analysing, which is used for getting rid of incorrectly segmented parts of the image and leaving the image area that corresponds the hands field.
3. Recognition of gestures—properties from the received image is reading. After this, gesture can be classified with the help of using a fuzzy logic conclusion system.
4. Commands execution on the basis of the recognized gesture.

Current societies in developed countries face a serious problem of aged population.

The growing number of people with reduced health and capabilities, allied with the fact that elders are reluctant to leave their own homes to move to nursing homes, requires innovative solutions since continuous home care can be very expensive and dedicated 24/7 care can only be accomplished by more than one care-giver.

So, it seems very attractive to develop a robotic platform for elderly care integrated in the Living Usability Lab for Next Generation Networks. Such project aims at developing technologies and services tailored to enable the active aging and independent living of the elderly population.

As the rapidly growing number of disabled people in the world, the role of electric wheelchair [24] becomes important to enhance the mobility for them.

Many researchers pay and will pay attention to how to solve the problems in daily life for the wheelchair users, rarely cares for their physical exercises.

And also very important scientific direction deals with using socially assistive human-robot interaction to motivate physical exercise for older adults [25].

Special attention should be paid to the design, implementation, and user study evaluation of a socially assistive robot system which can be designed to engage elderly users in physical exercise aimed at achieving health benefits and improving quality of life.

Many research efforts have been carried out for finding strategies for motivating people to exercise regularly.

It is known about an Android-based mobile application, called Everywhere Run [26], that aims at motivating and supporting people during their running activities, behaving as a virtual personal trainer.

Everywhere Run fosters the interaction between users and real personal trainers, in order to make it easy to non expert people to start working out in a healthy and safe way.

For elderly and or physically disabled people who have lost their body functioning of motions due to geriatric disorders, and/or disease processes including trauma, sports injuries, spinal cord injuries, occupational injuries, and strokes, it is necessary to continue in future a development and research of 3DOF mobile robotic exoskeletons [27] for rehabilitation and for assisting motion of elbow and shoulder, since human shoulder and elbow motions are involved in a lot of activities of everyday life.

3   The Role of Information Technology in Culture and Sport Development

Let me say a few words about information technologies perspectives with correlation to culture and sport.

The Internet, World Wide Web, and related information and communication technologies (ICTs), have rapidly spread to a large number of countries and cultures.

Many of these technologies facilitate and mediate activities whose modes and means bind closely to culture.

So, information and communication technology is advancing rapidly and the world community has undergone a fundamental changes.

Activities, jobs, skills, sports, cultures, needs and have changed and been affected by this technology.

Immigrants represent a substantial part of European society.

After emigration, they can suffer from fundamental changes in their socioeconomic environment.

Therefore, supportive ICT services (e.g. for language learning or job search) have high potential to ease inclusion [28], especially for newly arrived immigrants with low education.

A lot of publications [29, 30] are devoted to perspectives of digital libraries and librarian consortia and some publications consider the solving problem—how are on-line digital libraries changing theatre studies and memories.

Design and implementation of an "Web Application Programming Interface (API)" for the automatic translation from one language to another is very important research direction [31].

For example, API allows exposing the service of automatic translation between two Colombia's language pairs (Spanish and Wayuunaiki case) using a evolving development model based on prototypes and through solutions based on open source technologies [31].

This innovative development uses the Information Technology and Communication to focus on Aboriginal languages translation enhancing the integration of technology with communities and allowing inclusion of languages and cultures conserving in the ICT ecosystem in Colombia.

Some countries in the world are multi-ethnic nations with rich and diverse culture traditions.

State-wide standardized textbooks, however, take little into account the local ethnic minority cultures; thus limiting the opportunities of preserving the unique ethnic traditions through education [32–34].

On the other hand, property of information technology requires information technology teaching to combine knowledge with local culture.

Curriculum resource can bridge the gap between textbook knowledge and ethnic minority cultures.

So, it necessary to implement ideas on development of effective curriculum resources website based on ethnic minority cultures [32, 33] for use by teachers, and on reform of classroom teaching in order to incorporate ethnic minority cultures into education of information technology.

Digital museum has become an important mode in culture communication with the great development of information technology.

Specially, thematic digital museum mode [35] has great advantages in propagation of culture.

This mode would develop the resources and the educational functions of digital museum.

With the spurt progress of modern information technology, information technology has become the core of the advanced productive forces in modern science and technology in the information age.

And the development of information technology makes the cultural industry enter a new field [36].

It is necessary in future to explore the coupling relationship between cultural industry and information technology.

Some investigations in twenty-first century are devoted to [37–40]:

- Mobile technology application to study cultural heritage in the people's everyday life.
- Effect of culture, age, and language on quality of services and adoption of Internet Protocol (IP) applications.
- Examining influence of national culture on individuals' attitude and use of information and communication technology.
- Discovering cultural trails in the social media.
- Public library alliance based on the public cultural services system.
- Application of digital information technology in architectural designing field.
- Ethical aspects as "cyber ethics", "internet ethics" and "computer ethics" that involve using information technology and resources in higher education.
- Understanding information technology induced changes in culture and perspectives on cultural change within an integrated information systems environment.

- Cultural challenges in information systems innovation.
- Sports loading computer-aided system which can support coaches for scientific sports training better.
- Organic combination of multi-media information technology and college physical education teaching.
- Intervention of information technology in sports training and application of information technology in the field of sport science and sport management.
- The importance of information technology in the development of the school sports.
- Smart technology and technical architecture of Smart Sport, the key technologies application method in this architecture, including body sensor networks, Internet of Things, cloud computing and data mining.
- Design and development of large-scale sport events information system.
- Sport video systems development.
- The diverse information technologies that can be used to provide athletes with relevant feedback.
- Person tracking systems for sport applications that can track athletes during training and/or in competition.
- The design of customer relationship management system in sports enterprises.
- The research on sports game news information extraction and others.

4   Conclusions

The main conception of this presentation deals with analysis of perspectives of information science, robotics, automation, culture and sports for twenty-first century and with aspects of their correlation and inter-coupling.

Scientists from different countries, including Spain and Ukraine, conduct research in this very important scientific field.

For, example, this VIII International Act "Science, Culture and Sports in the 21st Century" confirms paying a great attention to discussed problem from the President and all members of Royal Academy of Economic and Financial Sciences of Spain and from the distinguished participants from Algeria, Azerbajan, Italy, Romania, Spain, Ukraine and other countries.

Among other examples:

- The International conference "Economics, Management and Optimization in Sports. After the Impact of the Financial Crisis" that took place in Barcelona, Spain on December 1–3, 2009.
- First Ukrainian-German conference "Informatics. Culture. Techniques", which was organized by Odessa National Polytechnic University and Berlin University of Applied Science and took place in Odessa, Ukraine on February 18–19, 2013.

A lot of articles and books [39, 41, 42] were published concerning the problem "Science, Culture and Sports". Among most important books in this field are:

1. Les Universitats En El Centenari Del Futbol Club Barcelona. Estudis En L'Ambit De L'Esport. Jaime Gil Aluja (Editor), Proleg, Josef Lluis Nunez, 1999.
2. Optimal Strategies in Sports Economics and Management. Sergiy Butenko, Jaime Gil-Lafuente, Panos M. Pardalos (Editors), Springer-Vertag, Heidelberg, Dordrecht, London, New York, 2010.
3. Kultur und Informatik: Visual Worlds & Interactive Spaces. Regina Franken-Wendelstorf, Elizabeth Lindinger, Jurgen Sieck (Editors), VertagWerner Hutsbusch, Gluckstadt, 2013.

No doubts that we should consolidate our scientific cooperation in the discussed field for mutual benefit and for sustainable development of our societies.

Acknowledgments I gratefully acknowledge the support of this research direction by Royal Academy of Economic and Financial Sciences (RACEF, Spain), Petro Mohyla Black Sea State University (PMBSSU, Ukraine), RACEF President Prof. Dr. Jaime Gil Aluja and PMBSSU President Prof. Dr. Leonid P. Klymenko.

References

1. M. Tang, L. Yang, H. Zhou, Applications and safety of nanotechnology and nanomaterials in sports. Lect. Notes Elect. Eng. **207** LNEE (2013)
2. C. Bartneck, T. Suzuki, T. Kanda, T. Nomura, The influence of people's culture and prior experiences with Aibo on their attitude towards robots. AI Soc. **21**(1) (2007)
3. T. Nomura, T. Kanda, T. Suzuki, J. Han, N. Shin, J. Burke, K. Kato, Implications on humanoid robots in pedagogical applications from cross-cultural analysis between Japan, Korea, and the USA. Proceedings - IEEE International Workshop on Robot and Human Interactive Communication, art. no. 4415237 (2007)
4. T.L. Manuel, Creating a robot culture: an interview with luc steels. IEEE Intell. Syst. **18**(3) (2003)
5. Y. Ukai, J.Rekimoto, Swimoid: A swim support system using an underwater buddy robot. ACM International Conference Proceeding Series (2013)
6. W.S. Erdmann, Problems of sport biomechanics and robotics. Int. J. Adv. Robotic Syst. **10** (2013)
7. Ł. Szweda, D. Wilusz, J. Flotyński, Application of NXT based robots for teaching Java-based concurrency. Lect. Notes Comput. Sci. (including subseries Lecture Notes in Artificial Intelligence and Lecture Notes in Bioinformatics) **7516** LNCS (2012)
8. P. Benavidez, C. Gleinser, A. Jaimes, J. Labrado, C. Riojas, M. Jamshidi, L.B. Endowed, Design of semi-autonomous robots for competitive robotics, World Automation Congress Proceedings, art. no. 6321276 (2012)
9. S. Nadarajah, K. Sundaraj, Wireless communication in robot soccer: A case study of existing technologies, 2012 I.E. Conference on Sustainable Utilization and Development in Engineering and Technology, STUDENT 2012 - Conference Booklet, art. no. 6408360 (2012)
10. M. Ogawa, K. Ikeuchi, Y. Sato, S. Kudoh, T. Tomizawa, T. Suehiro, S. Shimizu, Towards air hockey robot with tactics - Statistical analysis from measurement of eye movement, 2012 I.E. International Conference on Mechatronics and Automation, ICMA 2012, art. no. 6282343 (2012)

11. H. Su, D. Xu, G.-D. Chen, Z.-J. Fang, Striking position selection based on two-step multi-purpose fuzzy decision method for robotic table tennis. Kongzhi Lilun Yu Yingyong/Control Theory and Applications **30**(5) (2013)
12. H. Aoki, Y. Fujimoto, S. Suzuki, E. Sato-Shimokawara, T. Yamaguchi, Difference in physiolog ical responses by different cultural greetings using a robot, IWACIII 2011 - International Workshop on Advanced Computational Intelligence and Intelligent Informatics, Proceedings (2011)
13. G. Trovato, T. Kishi, N/Endo, K. Hashimoto, A. Takanishi, A cross-cultural study on genera tion of culture dependent facial expressions of humanoid social robot. Lect. Notes Comput. Sci. (including subseries Lecture Notes in Artificial Intelligence and Lecture Notes in Bioin formatics) **7621** LNAI (2012)
14. H.-P. Yueh, W. Lin, The interaction between human and the home service robot on a daily life cycle. Lect. Notes Comput. Sci. (including subseries Lecture Notes in Artificial Intelligence and Lecture Notes in Bioinformatics) **8024** LNCS (Part 2) (2013)
15. V. Evers, H.C. Maldonado, T.L. Brodecki, P.J. Hinds, Relational vs. group self-construal: Untangling the role of national culture in HRI, HRI 2008 - Proceedings of the 3rd ACM/IEEE International Conference on Human-Robot Interaction: Living with Robots (2008)
16. J. Zhen, H. Aoki, E. Sato-Shimokawara, T. Yamaguchi, Sino-japanese culture differences in interaction robot system, 2012 IEEE/SICE International Symposium on System Integration, SII 2012, art. no. 6427370 (2012)
17. Y.V. Vizilter, S.Y. Zheltov, V.A. Knyaz, A.N. Hodarev, A.V. Morzhin, *Processing and the analysis of digital images with examples on LabVIEW IMAQ Vision* (DMK Press, Moscow, 2007) (in Russian)
18. S. Cunningham, A. Chellali, I. Jaffre, J. Classe, C.G.L. Cao, Effects of experience and workplace culture in human-robot team interaction in robotic surgery: a case study. Int. J. Social Robotics **5**(1) (2013)
19. E. Akdogan, E. Tacgin, A.M. Adli, Knee Rehabilitation Using An Intelligent Robotic System, Intelligent Manufacturing, Springer Science+Business Media, LLC, **2** (20) (2009)
20. Y.P. Kondratenko, G.V. Kondratenko, Robotic system with myoelectric adapying for reabilitation tasks, acta of bioengineering and biomechanics, in Proceedings of the 17th International Conference BIOMECHANICS (Zakopane, Poland), **3**, Supplement 2 (2001)
21. R.S. Machmud-Zade, Y.P. Kondratenko, Myoelectrical robotic system for rehabilitation patients with restricted movement. J. Chongqing Univ. **12**, 2 (1989) (in Chinese)
22. A.S. Dotsenko, Y.P. Kondratenko, Computer hand gesture recognition using fuzzy logic, Pattern Recognition and Information Processing PRIP'2011, Proceedings of the 11th International Conference (18–20 May, 2011), Minsk, Belarus, (2011)
23. S.G. Antoshchuk, A.A. Poplavskij, E.V. Tkachenko, V.Y. Kondratenko, Statistics of the non-numerical data in models of preliminary processing of images. Radio-electronic Comput. Syst. **6**(33) (2008) (in Ukrainian)
24. S. Jia, J. Yan, J. Fan, X. Li, L. Gao, Multimodal intelligent wheelchair control based on fuzzy algorithm, 2012 I.E. International Conference on Information and Automation, ICIA 2012, art. no. 6246880 (2012)
25. J. Fasola, M.J. Matarić, Using socially assistive human-robot interaction to motivate physical exercise for older adults. Proceedings of the IEEE **100** (8), art. no. 6235980 (2012)
26. F. Mulas, S. Carta, P. Pilloni, M. Manca, Everywhere Run: A virtual personal trainer for supporting people in their running activity. ACM International Conference Proceeding Series, art. no. 70 (2011).
27. M. H. Rahman, K. Kiguchi, Md. M. Rahman, M. Sasaki, Robotic exoskeleton for rehabilitation and motion assist, 1st International Conference on Industrial and Information Systems, ICIIS 2006, art. no. 4155186 (2006)
28. J. Bobeth, S. Schreitter, S. Schmehl, S. Deutsch, M. Tscheligi, User-centered design between cultures: Designing for and with immigrants. Lect. Notes Comput. Sci. (including subseries Lecture Notes in Artificial Intelligence and Lecture Notes in Bioinformatics) **8120** LNCS (Part 4) (2013)

29. M. Giacobbe Borelli, How are on-line digital libraries changing theatre studies and memories? Lect. Notes Comput. Sci. (including subseries Lecture Notes in Artificial Intelligence and Lecture Notes in Bioinformatics) **7990** LNCS (2013)
30. L. Li, H. Xiao, Relational study on the public library alliance based on the public cultural services system. Frontiers of Energy and Environmental Engineering, Proceedings of the 2012 International Conference on Frontiers of Energy and Environmental Engineering, ICFEEE'2012 (2013)
31. D.I. Fernández, O.Q. Gamboa, J.M. Atencia, O.E.H. Bedoya, Design and implementation of an "Web API" for the automatic translation Colombia's language pairs: Spanish-Wayuunaiki case, 2013 I.E. Colombian Conference on Communications and Computing, COLCOM 2013 - Conference Proceedings , art. no. 6564817 (2013)
32. W. Jia, The importance of information technology in the development of the school sports, Proceedings - 2012 I.E. Symposium on Robotics and Applications, ISRA 2012, art. no. 6219257 (2012)
33. A. Kamali, D. Becker, L. Kianmehr, Deception, ethics, and information technology: Policy implications, Proceedings of the Information Systems Education Conference, **ISECON 29** (2012)
34. B. Zhao, F. Xiao, M.-K. Liu, Z. Jie, Y.-D. Lu, On development and design of information technology curriculum resource website based on ethnic minority cultures, Proceedings of the 8th International Conference on Computer Science and Education, ICCSE 2013, art. no. 6554093 (2013)
35. Y. Chang, G.-Y. Rong, Application of thematic digital museum in communications of ceramic culture, Proceedings of the 8th International Conference on Computer Science and Education, ICCSE 2013, art. no. 6554071 (2013)
36. Y. Zhao, The linkage logic and development of information technology and culture industry. Lect. Notes Elect. Eng. **163** LNEE (2013)
37. D.G. Liebermann, L. Katz, M.D. Hughes, R.M. Bartlett, J. McClements, I.M. Franks, Advances in the application of information technology to sport performance. J. Sports Sci. **20**(10) (2002)
38. R.O. Maghrabi, P. Palvia, Understanding information technology (IT) induced changes in culture, 18th Americas Conference on Information Systems 2012, AMCIS 2012 **3** (2012)
39. Sergiy Butenko, Jaime Gil-Lafuente, Panos M. Pardalos (eds.), *Optimal Strategies in Sports Economics and Management* (Springer-Verlag, Heidelberg, Dordrecht, London, New York, 2010)
40. F. Shen, J. Li, Z. Wang, Information technology and its application in sports science. Adv. Intell. Soft Comput. **159 AISC**(1) (2012)
41. Regina Franken-Wendelstorf, Elizabeth Lindinger, Jurgen Sieck (eds.), *Kultur und Informatik: Visual Worlds & Interactive Spaces* (VertagWerner Hutsbusch, Gluckstadt, 2013)
42. Jaime Gil Aluja (ed), *Les Universitats En El Centenari Del Futbol Club Barcelona* (Estudis En L'Ambit De L'Esport, Proleg, Josef Lluis Nunez, 1999)

Culture and Sports as Catalysts for the Twenty-First Century Society

Mario Aguer Hortal

Abstract The twenty-first century has infinite possibilities, as well as great uncertainties about the major opportunities and challenges facing humanity. The final combination of variables that will have an impact on the different dynamics is yet unknown. If present trends continue, the world of culture and sport will present a spectacular development. The increase of sport participation among large sections of the population, new products and multimedia channels, the emergence of new materials, new sports and new rules will revolutionize sports. Emerging economies will hold much of that demand and will strive to organize sports mega-events that will launch them onto the international arena. There will be new business opportunities and employment sites. Factor mobility will reach higher degrees in a globalized world. The rise of celebrities and global sport organizations will be another phenomenon, together with the resizing of competitions in new organizational levels. Sports will have a role in a culturally globalized world marked by signs of local identities that will not disappear. Culturally, we will become more global, but sportingly we will not quit being local.

Keywords Sport • Emerging economies • Social phenomenon • Consumer society

1 Introduction

It would be presumptuous of me to make predictions for the coming decades in the world of culture and sports, covering a large time span worldwide. The inevitable academic humility that comes from the scientific method, warns us about the danger

M.A. Hortal (✉)
Royal Academy of Economics and Finances, Barcelona, Spain
e-mail: ceurabcn@cerasa.es

A.M. Gil-Lafuente and C. Zopounidis (eds.), *Decision Making and Knowledge Decision Support Systems*, Lecture Notes in Economics and Mathematical Systems 675, DOI 10.1007/978-3-319-03907-7_7, © Springer International Publishing Switzerland 2015

of this useless task. However, we will try to reflect on a topic which is so familiar to most people, namely sports as a form of cultural manifestation.

The horizon presents us with infinite possibilities, combining visible and invisible elements. Unexpected events shatter the most rational visions. The prediction of landscapes is based on a fundamental ideal: we observe the past experience to extrapolate future with some corrections, in which case randomness is either not taken into account, or, in being too much contemplated, surprises us no more. There is no doubt that we fail more than we succeed.

2 Background Assumptions

We would like to highlight two variables and a determinant that will influence the study. On the one hand the population growth, and directly related to it, the progress of the international economy. The key determinant will be the magnitude of the disasters that will surely occur in the coming decades, as has been happening with different periodicity since the world began. We assume that globalization is the framing factor of the landscapes that we have now and will have in the future.

The evolution of the world's population, currently 7,000 millions, will be one of the key factors. In four or five decades the planet could be populated by 9,000 or 10,000 millions, of which, predictably, four-fifths would be living in Asia or Africa.

The international economy is tied to the population growth variable. May observed trends continue, the world in the twenty-first century, as a whole, will be more populated and richer. It goes without saying the increasing importance of emerging economies, the relative decline of the West as an economic engine and globalization force of markets. These are unquestionable aspects which have acted strongly from recent decades.

There is little doubt that the average productivity per unit of labor will tend to increase as markets and free trade develop. Wealth will improve and impoverished and depressed areas of the world will catch up with developed economies. There will be an increase in the population that can be fed, and a rise in health and life expectancy. Globally the working population will increase dramatically, as well as its consumption capacity. Predictably, economic differences among countries shall be reduced, maybe not among the individuals of each country.

That said we cannot ignore the determinant to which we made reference above: misfortunes. The magnitude of the disasters that will happen will affect the evolution of the two variables: population and economy. We do not Know the dimensions of future nuclear, civil or military accidents, the presence of international terrorism, the importance of the economic crisis, the size of the armed conflicts and their internationalization capacity, the spread of diseases, potential pandemics, and climate evolution, of weather phenomena and their impact on human life. These things will possibly happen, although their magnitude is unknown to us.

Having made such considerations, the present work will be based on a linear trend model, predictable and distant from catastrophic or naive extremes, but without ignoring the interdependence of our world and the unpredictability of foreseeing unexpected impacts of systemic variables. The different ingredients that are here presented will be combined and the relationships established between them will be show.

3 The Cultural Context of Sport

The culture of the twenty-first century, understood as the set of patterns or collective beliefs and the communications established among individuals in order to respond to specific needs is, to a certain extent, not as rich and plural as the twentieth century was, but it is more widespread among the world population, that shares a variety of aesthetic tastes and preferences in a highly standardized world market.

The cultural environment of individuals is crucial to understand their behavior. Sport is a fundamentally cultural as well as social phenomenon. Culture, to a certain extent, is the very basis that sets people in dense and complex social relationships. It is a body of visions, symbols, icons and values. We are social beings, because we are moldable. This is the starting point for a statement that few will question: the cultural dynamics of the twenty-first century will be dual. On the one hand, there will be a standardization and convergence to the homogenization of human behavior around the world—in food, clothing, education, consumption and lifestyle—, but on the other hand, individuals will be very preoccupied about of the preserving of their own heritage and the protection of their local culture, that one thing that makes them different, giving them a sense of belonging to a group, to a collective, to a community.

The world of sports will emulate this double reality. That is to say, even if we can buy a Nike T-shirt of the Brazilian soccer player Neymar Jr. in any big city in the world, we watch live, along with millions of other viewers, a match of the Copa America between the USA and New Zealand, or see the latest Usain Bolt in the 100 m sprint Olympic final, individuals will continue to try hard to preserve and maintain certain practices and local consumption. Predictably, local community insiders will have a major role in the twenty-first century. There will be a shared worldview arising from the market economy and mass culture, along with a variety of local cultures of different size and vivacity.

At this point, it is possible to establish a certain analogy between the world of sports and religion. They share faith and worship. I think the sports are, and will be in the future, a kind of local religion for millions of people. It is worth noticing that regular practice, but above all, participation in many sporting events involve, in many cases, rites and ceremonies: a particular liturgy, songs, badges, distinctive clothing, a deeply rooted sense of belonging, and veneration of sport icons of the past and present. This, in a way, reminds us of the religious cult. In many places, stadiums seem to have replaced the old cathedrals—contrary to what was seen just a few decades ago. This attempt of comparison points to a reality that should not be underestimated. The hallmarks still exercise a strong magnetic power in the essence of the person as a social being.

Still, mass culture will progressively become more widespread in the twenty-first century. Consumer society based on advertising will proliferate and extend its presence across the globe. This will generate the creation of products, ideas and dreams that will be characterized by its uniformity and standardization. The global village generates global, standardized and interchangeable products. Although the West will lose its protagonist role in its contribution to the production of mass culture, the American way of live will still be a model. Probably the basis of the American life-style and western life style as well, tailored to subtle aesthetic local features, will be reproduced in different cultural and sporting products occurring worldwide. The media will show interchangeable living styles. Culturally, we will be more global, but in the field of sports we will not quit being local.

4 Sport in the Twenty-First Century

Sport and its practice are social phenomena directly associated with play. Human beings, all of us, to a greater or lesser extent, like to play and compete. Over time, sports develop as much as the wealth of societies and welfare of individuals. The poor, wherever they live, do not often do sports. We are reminded of them in Abraham Maslow's hierarchy of human needs.

From this assumption, and expecting the world's wealth in the twenty-first century to rise significantly, there is one evident conclusion: spots practice and sports television event consumption of a variety of types will spread all around the globe. Jogging, basketball, fitness, swimming, football, korfball or cricket, for example, will become ordinary practice of millions of people who will demand equipment and associated goods and services.

The inclusion of school sports in less developed countries will also consolidate a decisive step in changing its practice and consumption into an economic engine. It cannot go unnoticed that sports offer great appeal as training and ethical values. The values of effort, sacrifice, teamwork and ambition to improve are perfect matches to the prevailing model of an increasingly globalized culture.

Interest in health, in the cult of the body, or in the leisure activities associated to it, will cause a major increase in the practice of sports with an increased in the demand for goods and services. The sector will probably be a good employment niche in the coming decades: direct and indirect human resources, equipment, maintenance, services, sports medicine, engineering and applied services, rehabilitation, etc., will have an extensive demand in the market.

With globalization as a backdrop, the mobility of factors linked to sports will have identical behavior to what has now become common place in other sectors of the economy, although that seemed unthinkable just a few years ago, or limited to a small global elite of athletes in mass spectator sports. It is likely we will progressively witness a greater geographic mobility of players, experts, coaches, technicians and professionals of all kinds. The most economically advanced countries holding more extensive sporting tradition may become exporters of certain human assets, while emerging countries of others. In any case, the dynamics between the different geographical areas of the world will be rich and multidirectional.

As the present century society progressively becomes more urban, a growing appreciation of the quality of life will gradually develop, although with different intensity in different regions of the planet, counteracting a sedentary lifestyle, as well as the stress and isolation of individuals in urban contexts. This will have a positive impact on the business opportunities that may arise to meet growing demand expectations, with a positive impact on economic development and the labor market.

During the twenty-first century, something similar to what has been happening in the last 40 years in the West will happen in the rest of the world. Doing sports will become an ordinary practice for the majority of the population. Predictably, three variables related to economic development will particularly contribute to this: high concentration of population in the cities, higher rates of sedentary work and weight gain in the population. We shall not forget that the population of the twenty-first century will surely be, on the whole, more urban, older and more obese. The World Health Organization warns that as wealth expands so does the population overweight. All this helps sports to end up being more present in the lives of citizens.

Major world sporting events will increase their presence in the media. In a global village without borders, mass consumption of televised sports will increase. We will not only witness the proliferation of television channels specialized in all kinds of sports, sports as a social phenomenon will have a strong presence in the lives of people. International organizations and clubs negotiating broadcasting rights for certain sporting events will adjust their service start time schedules to where they hold their target audiences. Some figures about it. By 2050 the African population will triple to the European and 40 % of the total world population will live and consume in Asia. The products will adapt to this reality, times and customs.

No doubt there will be many changes in the sports world in the coming decades. Some representative areas will be: organizational levels, new materials, new game rules, media celebrities, and the mega sports events.

4.1 Organizational Levels

When discussing organizational levels, it is likely that traditional competitions of certain mass sports will be resized in favor of competitions in continental or sub-continental leagues. There will be a natural selection of sporting clubs in the same way that television rights will be concentrated in the hands of a few, who will be particularly interested in the attraction of great sports celebrities and in the expanding of their audiences.

It would not be surprising to find a European soccer league, for example, that marginalized or replaced the Champions League and relativized each national league. It will possibly exist two or more types of professional or semi-professional competitions. On the one hand, those with media exposure, which will increase geographic scales to gain visibility, audiences and resources; on the other hand, those maintaining a local audience, with little or no part in the sports business. A network of high-level interregional and intercontinental sports men will be present in the media throughout the twenty-first century.

In all this evolution we will see a modernization of the old traditional sports structures and a growing professionalization.

In the future, clubs in general will increasingly depend on private income, basically: shareholder funds, ticket sales—in case there is plentiful demand—and the greater or lesser amount of publicity they can attract. The rules of economic game will make clubs that are not competitive, quite simply disappear.

Perhaps many organizational levels of mass sports will downsize, not exactly for sporting limitations, but for economic reasons. In addition, large sports complexes, large stadiums will no longer be exclusively a space for sports. They will become multipurpose spaces with restaurants, shops, and will offer a range of leisure activities to attract new audiences from their brand image.

4.2 New Materials

New materials are another aspect to be taken into account. They will revolutionize the sport in the twenty-first century. A whole industry dedicated to changing technological advances into higher competitiveness will have a direct relation with motor sports, sailing, cycling and other sports. Aerodynamics, new fabrics and new training practices will revolutionize the sport.

The new materials, biological or mechanical—embedded in the human body—will contribute to the emergence of an ethical debate around its limits. The South African athlete Oscar Pistorius, a fleeting myth of world athletics who had two prostheses in his lower extremities, was the indirect inducer of this issue at the beginning of twenty-first century. A sadder example was the cyclist Lance Armstrong's. He was indicted on charges of systematic doping and subsequently withdrawn of seven Tours irregularly won. The lust for triumph and glory has been staining the ethical limits of competition and fair play, keeping the threat of a "cheatty" version of the Prometheus myth around. That greed and desire for sporting glory at any cost will not disappear. It will drag fools and scammers resorting to prohibited methods to wrong paths. Doping battle cannot stop.

I would like to make reference to a theory stated by Eudald Carbonell and Robert Hall, the prestigious anthropologists of the archaeological site of Atapuerca, when, in their book Human Planet, they remind us of the tendency of our species to full mechanization, taking us in the direction of the creation of technobeings. In the twenty-first century, predictably, this tendency will be accentuated as the "creation" of post-humans through genetic engineering and human-computer symbiosis develops. The sports world will not be alien to this process.

Sports for the disabled will be one of the grandest beneficiaries of the advances that are expected to occur in the future. Paralympic sports, in one way or another, will attract more media attention and will be more competitive. The triumph of athletes will probably be more relevant than finding a feeling of integration, which may lose intensity. There is no doubt that engineering and prototype designs and prosthetics have a great potential in advancing these sports. It will increase the autonomy of athletes with disabilities of all kinds through devices that emit vibrations, sound or olfactory stimuli to guide them and help keep them on track during the competition. New bionic devices, through mechanical prostheses, will receive and understand commands from the brain, changing them into electrical impulses. This will revolutionize the world of disability and the paralympic sports.

4.3   New Game Rules

Many sports will have their rules revised to make their practice more attractive, but, above all, to make them more suitable for television. In order to increase audiences and find a niche in the fierce competition among sports, it is likely that corporate sponsors, television, websites and all platforms make changing the rules a faster process than it has been so far.

The ultimate goal is none other than make sports events more spectacular and attractive. Attracting and retaining the audiences. New sports will be created, not only risky but also strange ones, unique and unusual, as it was the case, for example, of korfball in its infancy. The new sports will gain prominence and increase their audiences. It would not be unusual that in future Olympic Games featured sports in unusual contexts, such as the air. Maybe with aerial ballets or something like that. Nor would it be surprising that urban sports practices related to skateboards or acrobatic MX bicycles gained greater presence in the regulated sport in the twenty-first century.

The essence of the philosophy of sport synthesized in the Olympic motto citius, altius, fortius, faster, higher and stronger, will gain a different approach in the twenty-first century. Possibly the new challenges of sports and athletes will be making sports more risky or daring, more surprising or extraordinary, and more technological.

Many clubs and stadiums will possibly generate new strategies to attract, retain and increase stadium attendance and ticket sales. Everything seems to indicate that digital and technological advances will be incorporated in order to enhance the experience of seeing a live meeting. The audience and players may come to be connected unidirectionally. Perhaps we will hear them as they speak and feel their heartbeat or listen to their comments on the field. Their shirts may incorporate tiny sensors that will possibly transmit all kinds of information. The aim is making the experience of watching a game on the field more to realistic and vivid. It is very likely that in the twenty-first century the conventional relationship between spectators and players will be deeply changed.

4.4   The Celebrities

The big clubs and big sports celebrities will seek for greater global projection. They will gain the status of real media icons who will transcend sports phenomenon to, directly or covertly, become phenomena of mass consumption in a plain business of a highly competitive global market. Sports merchandising associated with media celebrities and sports clubs will flood even more, supermarkets and shops worldwide.

Global sports stars will probably become more ephemeral than their successors. Sports stardom will last less, because there will be more stars that shine more and more competition for being in the spotlight. Nothing to do with the austere athlete profile that prevailed until the late twentieth century. Millions of people will identify with the new icons that will become objects of desire for the world of advertising and mass consumption.

4.5   Spots Mega-Events

As a result of the eclosion of the sports world and media, such as has been indicated by authors like García Ferrando, we will witness the development of a new phenomenon: the growing importance of sports mega-events.

A sports mega-event is a large-scale event with an impressive global projection. In referring to this type of events, we do not only want to make reference to refer to the Olympics, also the to the Mediterranean Games, the Commonwealth Games, the America's Cup sailing yacht competitions, the Formula I and motorcycling races, the Grand Slam Tennis Tournament, the world cup or the rugby leagues and continental and world basketball cups, baseball or football.

Likewise, Ramón Llopis, believes that mega sporting events are prominent elements of globalized societies of late modernity, while central to the new economies of signs and space. The attractiveness of these events has not ceased growing in recent times and it is possible that the trend is on the rise. The fact that the Olympics is watched by 4,000 or 5,000 million people is an indicator of its magnitude.

Emerging economies will compete more than anyone else to get to organize mega-events that help them be propelled and positioned in the global arena. These events create value for the host economies. From an economic and tourism perspective, they regenerate the urban fabric, streamline areas with limited visibility, help correct inequalities and reinforce the nationalist sentiment of the people, their pride and national dignity.

In the twenty-first century many cities in the new economies will struggle to hold world renowned sporting events. Being a host implies prestigious international projection, it means an opportunity to modernize equipment and infrastructure, and it has expected positive results. The sport will remain a most coveted instrument in the political struggle.

5   Conclusion

The sport has been gaining presence in our lives. It will continue to do and more intensely, either through popular sport, federated or associated, or through what is being called fitness. More people, of all ages, will do sports. Regular physical activity, in order to improve physical fitness will promote a grand development of sports centers based on an extensive range of spaces for sporting practices, both free and directed ones.

The idea of sport for all will increasingly become a reality within the reach of most people. Clubs and gyms belonging to different holders will be combined with interventions on urban and green areas, where through different programs and devices will promote free sport. The media will also multiply its sports multimedia products in unimaginable ways now. New employment sectors and business opportunities will be directly related to all this.

New countries, new economies will concentrate a large part of the demands of sports products in the world and will struggle to organize mega-sports events in order to gain prominence in the international arena.

New materials, new rules and new sports will change the sports landscape. The paralympics will become more prestigious. The debate around the limits of the machine-man, the fair play, or the shadow of doping will not abandon the world of sports.

The sport will continue to be a major tool for integration in the twenty-first century. However, certain weaknesses threaten the world of sport. It is worth noting that in the trend that has been observed when analyzing maps of sport practice, certain groups continue to be underrepresented in the sports family. I mean, women, disabled persons, the elderly and the disadvantaged. This under-representation is not only limited to sport practice, it also influences the organization leader ratios, regional, national and international associations and federations, or the degree of professionalism, visibility and public recognition.

The impact of globalization blurs borders and reduces cultural differences, but on the other hand, human societies are basically local and as Zygmunt Bauman tells us in his extensive bibliography on liquid modernity, the individuals who compose it become progressively more fragile, in need of references and are more vulnerable and lonely.

Finally, it is relevant to acknowledge that the influence of sport in universal culture in the twenty-first century opens up grand opportunities. Long-term forecasts, besides made on a global scale, are only approximate and subject to arbitrary events which are impossible to predict. Still, we are optimistic about the potential of human, economic and social development that sports can generate the in this century.

References

1. Z. Baumn, *La sociedad líquida* (Paidos, Barcelona, 2000)
2. E. Carbonell, R. Sala, *Planeta humano* (Península, Barcelona, 2000)
3. P. Carolina, Aranzadi Thomson Reuters, Pamplona (2012)
4. L. Cazorla, Thomson Reuters Aranzadi, Pamplona (2013)
5. M. Ferrando, Eur. J. Sport Soc. **7**, (2010)
6. Fondo Monetario Internacional, *World economic outlook. Database* (Fondo Monetario Internacional, Washington, 2012)
7. M. Godet, *Creating futures: Scenario planning as a strategic management tool* (Economica, Paris, 2006)
8. C. Hakim, *El capital erótico, Debate*, Madrid (2012)
9. R. Llopis, *Megaeventos deportivos. Perspectivas científicas y estudios de caso* (Editorial UOC, Barcelona, 2012)
10. J. Marin, Deporte, comunicación y cultura. Comunicación Social (2012)
11. J. Pérez, *Ética y deporte* (Desclée de Brouwer, Bilbao, 2011)
12. F. Raventós, *Un futur incert. Economia, geopolítica i governança mundial en el segle XXI* (La Magrana, Barcelona, 2012)
13. J. Requeijo, *Odisea 2050. La economía mundial del siglo XXI* (Alianza Editorial, Madrid, 2009)
14. The Economist, *El mundo en 2050. Todas las tendencias que cambiarán el planeta. Gestión 2000, Barcelona* (2013)
15. VV. AA, *Global governance 2025: at a critical juncture*. CIA. Estats Units (2010)

The Forgotten Effects Model on Selection Policies to Climate Change Adaptation

Anna Maria Gil-Lafuente and Jaime Alexander López-Guauque

Abstract The impacts of climate change have become a growing question for the various players in the field. Adapting to such impacts is seen as an uncertain and complex phenomenon the more one gets involved in each of the key processes. Therefore, the degree of involvement or impact will have a larger or lesser effect, on an uncertain scale, in a given a period of time, and under homogenous conditions. Several elements of success in the decision-making are analyzed and discussed. Such elements, with different weightings, are implicit in the decision-making processes. From the forgotten effects theory, we can establish the accumulated effects of first and second generation in order to determine multiplier or enhancer effects—where our efforts will be focused. This will allow a longer scope for decisions to be made and strategies to be formulated as well as for the valuation of various groups and scales in which adaptations will be evaluated.

Keywords Climate change • Adaptation • Forgotten effects model • Decision making

A.M. Gil-Lafuente (✉)
Faculty of Economics and Business, University of Barcelona, Barcelona, Spain
e-mail: amgil@ub.edu

J.A. López-Guauque
Faculty of Economics and Business, University of Barcelona, Diagonal, 690, Barcelona 08034, Spain

A.M. Gil-Lafuente and C. Zopounidis (eds.), *Decision Making and Knowledge Decision Support Systems*, Lecture Notes in Economics and Mathematical Systems 675, DOI 10.1007/978-3-319-03907-7_8, © Springer International Publishing Switzerland 2015

1   Introduction

We are in a time of great change in all areas. A time, in which man is leaving his footprint, causing irreversible climate changes worldwide, according to scientific sources. Climate change is responsible for the increase in the frequency and severity of extreme weather events in all aspects of nature.

Consequently, the economic losses caused by natural disasters could increase significantly in the coming years. This will have considerable negative effects on the various socio-economic actors worldwide. On the one hand, an increased risk of extreme weather requires a reassessment of the expected changes in the damage and the inclusion of an appropriate projection of climate change in risk management. Furthermore, the establishment and quantification of the added value can be generated in the formulation of globalized economic strategies and social solutions.

Disaster-related losses have been significant in the last decade, which poses significant challenges to the insurance industry worldwide. For example, in the last two decades (1991–2010) the United States experienced the second most damaging hurricane season in a century quantified in terms of damage normalized by inflation and wealth. In the decade 1926–1935, there was more damaged with higher costs due to hurricanes [1]. Referring to the climate in Europe, there have been substantial losses. Floods in Germany during the year 2002 caused losses of about €9.2 billion [2]. England has experienced two major floods in the summer of 2007, caused by extreme rainfall. Overall economic losses amounted to about four billion dollars, a catastrophic event, of which three million were secured. The magnitude of the economic losses suffered highlights the vulnerability of modern societies to changes in climate.

According to projections of the Intergovernmental Panel on Climate Change [3], it is noted that there may be an increase in the frequency and severity of extreme weather events such as rain (extreme), tropical cyclones and heat waves in some regions. The effects of climate change on small extreme weather events such as lightning and hail, remain uncertain [3]. The global warming potential of increasing vulnerability to extreme weather events is especially relevant to the insurance industry [4]. The insurance industry is the largest industry in the world in terms of revenue and insurers bear a large part of climate risks, such as damage caused by floods and storms [5].

Insurers argue that the settings of premiums and coverage levels are enough to adapt to changes in the patterns of loss. In fact, the flexible nature of the industry that is characterized by mostly short term contracts, allow fairly quick adjustment of premiums, ensuring their resilience against climate change. However, the lack of interest can be problematic, meaning greater exposure and insufficient loss and considerable decrease in the incorporation of premiums and risk management practices. The premium adjustments based on the experiences of previous claims may insufficiently reflect changes in the paradigm of the calculations of the probability of occurrence of extreme weather events due to its low loss situation.

The relevance of climate risks for the insurance sector is evident by observing past trends in insured economic losses and other natural disaster types. Data from past events relating to natural catastrophe losses collected by Munich Re [2], indicate the increase in global trends losses are already observable. The main factor behind this increase in losses has been social change. Continued economic growth, economic development and population growth, especially in vulnerable regions (e.g., coastal areas), combined with rapid extreme climate change, makes it possible to increase, acceleration and magnification rate and the size of the damage. The influence of climate change on trends in catastrophe losses is likely to be exacerbated by a large change in extreme climates, therefore, expected occurrence due to a change in average climatic conditions [6]. The best strategy for insurers seems to incorporate not only the expected changes in the likelihood of extreme weather conditions in the exposure assessment, pricing and management of risk, but perform a rethinking of models, ushering in capable tools, the treatment of uncertainty, getting hybrid based algorithms which involve stochastic data and subjective information. Projected changes in the likelihood of extreme weather events can be obtained from regional climate models, in which Royal Dutch Meteorological Institute (KNMI) has developed climate scenarios for the Netherlands [7]. There is a clear need for adaptation measures to reduce risk exposure given the historical emissions of greenhouse gases and its consequent effect on radioactive forcing in the future [1]. Furthermore, the evolution of society such as increased economic development in vulnerable areas such as coastal areas require the completion of risk reduction policies in order to guarantee the insurability of climate risks. Along with traditional measures commonly used to limit the risks, such as increased premiums and limited coverage, the insurance industry could play an important role in stimulating and promoting climate change mitigation and adaptation [5].

Climate change projections for the various countries mainly indicate an increased risk of extreme weather events [7]. However, the consequences of climate change for insurers not only are negative. For example, the probability of frost may decrease in the future which could reduce claims on certain crop insurance policies. Climate change and things, we can introduce new profitable business opportunities. Demand for insurance products today can increase and market new insurance contracts, non-existent so far, due to the changes that lie ahead as the economic losses due to climate change will increase the risks [8]. However, the problem with the insurability of risks due to weather can hinder the development of markets, primarily because of the inherent correlation of these risks associated with the likelihood and impact of extreme weather events [9]. Various associations worldwide have raised promising solutions to meet demand and compensate for weather-related damages that are not currently covered by private insurance, such as the risk of drought, floods, among others.

This paper will review several elements of success in the decision-making. Such elements are implicit in the decision-making processes.

The formulation of strategies in such uncertain scenarios greatly hinders the application of numerical methods. We propose the use of techniques arising from the theory of fuzzy subsets [10]. We know of the implementation of these theories in many scientific fields by various authors and applications to various fields of knowledge [11–22].

Given the forgotten effects theory developed by teachers Kaufmann and Gil Aluja [23], which allows a fuzzy logic approach, identifying the contribution of the causal relationships between adaptation strategies to climate change (variables), on the one hand, and between the guiding principles and indirect relationships with each of the actors on the other, provides important information in the decision making process for the design and improvement of the adaptive strategies and management of natural risks, both to countries and companies.

This paper aims to highlight the direct relationships between the guiding principles. The importance of each guiding principle depends on the context they are applied to, in particular regarding the stages in the adaptation process, the level of decision-making or the specific regional conditions. Also, indirect relations for each of the adaptation strategies, in response to climate change outlined variables.

The work is divided after the introduction—a section which summarizes the general characteristics that identify the basics of adaptation on climate change and decision-making under uncertainty. The following section discusses the forgotten effects. The next section outlines the methodological approach of the forgotten effects model. The next section presents, in summary, the results of the model. Finally, the conclusions are presented highlighting the values of the model proposed and implemented, showing how it contributes to the improvement of adaptive strategies management.

2   Adaptation on Climate Change

At the 2010 climate conference in Cancun, Mexico, parties to the United Nations Framework Convention on Climate Change (UNFCCC) affirmed that adaptation must be placed on a par with climate change mitigation. In practice, adaptation is climate-resilient development and natural resources management. In recent years, adaptation has emerged as a top priority on the international development agenda [24].

The question of evaluating and defining success of climate change adaptation activities has been widely discussed throughout the international evaluation community. Adaptation presents a range of uncertainties that makes evaluation particularly challenging. Despite, the uncertainties that plague climate change adaptation efforts, successful action is possible.

Evaluating climate change is made particularly difficult due to the uncertainty surrounding not only the climate change science, but also the overall scenario in which climate change adaptation activities are implemented. Climate change activities are presented by a multitude of potentially shifting baselines. Adaptation activities are constantly threatened by potential changes in economic, social, and other environmental conditions. The adaptation is still in its early stages however, it provides dimensions for assessing the likelihood of success of adaptation activities.

As we have seen, a large body of work is not totally directed at adaptation to specific climate change impacts but also focused on addressing the root causes of vulnerability. That we find many adaptation efforts in this middle ground of the continuum reflects the main challenge of climate change: learning to live with new sources of uncertainty. Given the relatively high uncertainty associated with the effects of climate change in many places, adaptation efforts can not universally focus on the planning of a new climate expected. Instead, actors reduce vulnerability to an uncertain climate for selected integrated strategies that can be assigned to a number of potential future climate change, and reducing vulnerability to climatic sources of damage.

Over time, as understanding improves climate risks, adaptation experience grows, and the effects of climate change are felt more strongly in more places, impacts oriented approaches, especially climate risk management approaches appear to reach be more widely implemented. However, the effectiveness of climate risk management depends largely on the ability to reduce the uncertainties surrounding climate impacts to a level at which the risk management tools can be applied reliably.

In some sectors and locations, this reduction of uncertainty will happen relatively quickly, while others may occur only in recent decades, or not at all. It is essential, therefore, that certainty about climate risk becoming a prerequisite for action on adaptation. Many of the places and the most vulnerable communities will not be able to address climate risks in the normal sense of risk management, for them, the task of adapting the core in place to strengthen the ability to cope with uncertainty.

However, the availability of good climate risk information does not necessarily mean that adaptation decisions easier or better.

3   Forgotten Effects

In all natural sequential processes, it is usual to ignore voluntarily or involuntarily some stage. Each result has forgotten side effects ranging repercussions throughout the web of incidence in a kind of combinatorial process. The incidence is a highly subjective concept, usually difficult to measure, but the analysis improves reasoned action and decision-making. Always and at all moments, mistakes and oversights have been made as a result of forgetfulness or neglect.

When the sequences of incidents, inferences or consequences appear, they are usually not treated properly. Often only two or three steps of reasoning are treated. In order to improve the investigation of forgotten effects or oversights computers and adequate mathematical models are used.

These models are usually the graphs with binary values in the arcs or vertices; these graphs can be valued by numbers in [0;1] or by intervals of confidence of [0;1]. Semantic judgments to unify and have a single evaluation criterion of the truth of the incident are used [23]. The Boolean theory, the theory of fuzzy subsets or the "expertons" are used.

The importance of the effects of the second, third, etc. generation is appreciable in all areas of decision making such as political, economic, business, medicine, biology, etc. [25]. Qualitative incidence matrices are used to investigate different mechanisms of cause/ effect that it is not yet possible to find out with the help of intuition or experience; it is possible to create new mechanisms between different sectors in order to transform certain situations [26].

The idea of the effects of the second generation in the socio-economic and ecological fields has been developed by Jean Fourastié. There are the effects that have not been considered during decision making process. These "forgotten effects" become obvious later. It is convenient to detect them in advance and take necessary measures a priori. Using this concept Fourastié developed a method that allows obtain forgotten effects during the decision making process.

Recently, other authors that have dedicated their research works to the theme of the forgotten effects: [23, 27, 28].

4   Methodology

The objective of this study is to obtain the causal relationships between the strategies and principles of climate change adaptation; rated these as success factors. This study shows the different views of the experts relating to the adaptation policies and the actions that influence its basic principles. Creating the cause-effect relations between the principles and strategies, it was obtained not only an overview of the adaptation response measures, with a more nuanced and broader approaches, but also the "forgotten effects" that were significant for the analysis but have not been taken into account.

Each of the variables proposed in this research have been carefully derived. Other studies have been made available to select variables [29].

The guiding principles[1] of climate change (subset A), which are **causes** for this case, are the following:

C1. Initiate adaptation, ensure commitment and management.
C2. Build knowledge and awareness.
C3. Identify and cooperate with relevant stakeholders.
C4. Work with uncertainties.
C5. Explore potential climate change impacts and vulnerabilities and identify priority concerns.
C6. Explore a wide spectrum of adaptation options.
C7. Prioritise adaptation options.
C8. Modify existing policies, structures and processes

---

[1] Source: [29] The following ten guiding principles are strongly interlinked and should be understood in an integrated way. The importance of each guiding principle depends on the context they are applied to, in particular regarding the stages in the adaptation process, the level of decision-making or the specific regional conditions.

Table 1 The adaptation strategies of climate change (subset B), which are effects for this case

Strategy	Description	Number of cases
E1—Changing natural resource management practices	Emphasizes new or different natural resource management practices (e.g., for managing water, land, protected areas, fisheries) as adaptation strategies.	57
E2—Building institutions	Creates new or strengthens existing institutions (e.g., establishing committees, identifying mechanisms for sharing information across institutional boundaries, training staff responsible for policy development)	43
E3—Launching planning processes	Sets in motion a specific process for adaptation planning (e.g., developing a disaster preparedness plan, convening stakeholders around vulnerability assessment findings)	35
E4—Raising awareness	Raises stakeholder awareness of climate change, specific climate impacts, adaptation strategies, or the environment in general	33
E5—Promoting technology change	Promotes implementation or development of a technology new to the location (e.g., irrigation technology, communications technology)	31
E6—Establishing monitoring/ early warning systems	Emphasizes the importance of creating, implementing, and/or maintaining monitoring and/or early warning systems	25
E7—Changing agricultural practices	Focuses on new or different agricultural practices as adaptation strategies	23
E8—Empowering people	Emphasizes literacy, gender empowerment, or the creation of income generation opportunities as a basis for adaptation	22
E9—Promoting policy change	Promotes establishing a new policy or adjusting an existing policy	14
E10—Improving infrastructure	Focuses on creating or improving built infrastructure (e.g., roads, sea walls, irrigation systems)	13
E11—Providing insurance mechanisms	Creates, modifies, or plans an insurance scheme	4
E12—Other strategies	Consists of three instances of relief work and one focused on eradication of climate-related diseases	4

Source: Adapted from [30]

C9. Avoid mal adaptation.
C10. Monitor and evaluate systematically.

The adaptation strategies[2] [30] of climate change (subset B), which are effects for this case, are shown in Table 1.

The analysis of the relations between each cause and effect gives a 10x12 rectangular matrix M where the cells show the direct impact of each principle

---

[2] Adaptation strategies employed in the 135 cases examined.

Table 2 A rectangular matrix M

$\left(\underset{\sim}{M}\right)$	E_1	E_2	E_3	E_4	E_5	E_6	E_7	E_8	E_9	E_{10}	E_{11}	E_{12}
C_1	0.7	0.6	0.8	0.6	0.2	0.8	0.6	0.7	0.6	0	0.7	0.6
C_2	0	0.4	0.5	0.4	0.7	0	0.4	0.9	0.6	0.1	0.3	0.8
C_3	0.8	0.5	0.7	0.8	0.7	0.8	0.2	0.8	0.6	0.1	0.2	0.8
C_4	0.9	0.3	0.8	0	0.1	0.6	0.1	0	0.1	0.1	0	0.2
C_5	0.8	0.2	0.8	0.2	0.5	0.9	0.5	0.7	0.5	0.2	0.3	0.8
C_6	0.7	0.8	0.7	0.5	0.6	0.8	0.4	0.8	0.7	0.7	0.4	0.8
C_7	0.6	0	0.8	0.4	0.6	0.9	0.4	0.6	0.4	0.7	0.1	0.7
C_8	0.8	0.7	0.9	0.4	0.8	0.4	0.5	0.6	0.9	0.9	0	0.3
C_9	0.8	0.5	0	0.6	0.4	0.5	0.7	0	0.6	0.4	0	0.3
C_{10}	0.5	0.7	0.7	0.4	0	0.5	0.7	0.1	0.2	0.2	0	0.6

on each strategy. The valuations are made using an endecadario system $\{0;0,1;\ldots0,7;\ldots;1\}$. The results are represented in Table 2.

Let's determine the effect that a cause produces on itself and through another relation effect/ cause, so that in all effects will be accumulated a result of a direct cause and a result produced through an indirect way.

A square matrix $\underset{\sim}{A}$ is made where the "causes" are placed as the rows and columns. The causes represent the principles that have to been done to participate in adaptation process. The valuation is made in [0;1].

A new square matrix $\underset{\sim}{B}$ is made by means of fuzzy cause-effect relations where both rows and columns represent "effects" taken from the matrix $\underset{\sim}{M}$.

Basing on the fuzzy matrixes of incidence such as $\underset{\sim}{M}$, $\underset{\sim}{A}$ and $\underset{\sim}{B}$ and in order to get accumulated effects of the 1 and 2 generation a new fuzzy matrix of incidence $\underset{\sim}{M^*} = \underset{\sim}{A} \circ \underset{\sim}{M} \circ \underset{\sim}{B}$ is obtained by maximin convolution between $\underset{\sim}{M}$, $\underset{\sim}{A}$ and $\underset{\sim}{B}$. The results are represented in Table 3.

To find the effects of the 2 generation let's separate from accumulated effects of $\underset{\sim}{M^*}$ the direct effects given in $\underset{\sim}{M}$. In this case a simple algebraic difference $\underset{\sim}{M^*} - \underset{\sim}{M}$ is made. See Table 4.

These cells are the following: (1;10) the effect of "Initiate adaptation, ensure commitment and management" on "Improving infrastructure" with 0.8; (2;1) the effect of "Build knowledge and awareness" on "Changing natural resource management practices" with 0.8; (2;6) the effect of "Build knowledge and awareness" on "Establishing monitoring/early warning systems" with 0.8; (8;11) the effect of "Modify existing policies, structures and processes" on "Providing insurance mechanisms" with 0.8; (10;5) the effect of "Monitor and evaluate systematically" on "Promoting technology change" with 0.8; (10;11) the effect of "Monitor and evaluate systematically" on "Providing insurance mechanisms" with 0.8.

These causal relations have not been taken into account, it is the case for the cells with 0.8 level or have been considered too weak, the case of the cells with 0.7 level.

Table 3 A fuzzy matrix of incidence M^*

M^*	E_1	E_2	E_3	E_4	E_5	E_6	E_7	E_8	E_9	E_{10}	E_{11}	E_{12}
C_1	0.7	0.7	0.8	0.7	0.8	0.8	0.7	0.7	0.7	0.8	0.7	0.7
C_2	0.8	0.7	0.7	0.8	0.7	0.8	0.7	0.9	0.7	0.7	0.8	0.8
C_3	0.8	0.7	0.8	0.8	0.8	0.8	0.7	0.8	0.7	0.8	0.8	0.8
C_4	0.9	0.7	0.8	0.7	0.7	0.7	0.6	0.7	0.7	0.8	0.7	0.7
C_5	0.8	0.8	0.8	0.8	0.8	0.9	0.6	0.8	0.8	0.8	0.8	0.8
C_6	0.8	0.8	0.8	0.8	0.8	0.9	0.7	0.9	0.8	0.8	0.8	0.8
C_7	0.8	0.7	0.8	0.8	0.8	0.9	0.7	0.8	0.7	0.8	0.8	0.8
C_8	0.8	0.8	0.9	0.8	0.8	0.8	0.7	0.8	0.9	0.9	0.8	0.8
C_9	0.8	0.7	0.7	0.7	0.7	0.7	0.7	0.7	0.7	0.7	0.7	0.7
C_{10}	0.8	0.7	0.8	0.8	0.8	0.8	0.7	0.7	0.7	0.8	0.8	0.8

Table 4 A fuzzy matrix of incidence $M^* - M$

	E_1	E_2	E_3	E_4	E_5	E_6	E_7	E_8	E_9	E_{10}	E_{11}	E_{12}
C_1	0	0.1	0	0.1	0.6	0	0.1	0	0.1	0.8	0	0.1
C_2	0.8	0.3	0.2	0.4	0	0.8	0.3	0	0.1	0.6	0.5	0
C_3	0	0.2	0.1	0	0.1	0	0.5	0	0.1	0.7	0.6	0
C_4	0	0.4	0	0.7	0.6	0.1	0.5	0.7	0.6	0.7	0.7	0.5
C_5	0	0.6	0	0.6	0.3	0	0.1	0.1	0.3	0.6	0.5	0
C_6	0.1	0	0.1	0.3	0.2	0.1	0.3	0.1	0.1	0.1	0.4	0
C_7	0.2	0.7	0	0.4	0.2	0	0.3	0.2	0.3	0.1	0.7	0.1
C_8	0	0.1	0	0.4	0	0.4	0.2	0.2	0	0	0.8	0.5
C_9	0	0.2	0.7	0.1	0.3	0.2	0	0.7	0.1	0.3	0.7	0.4
C_{10}	0.3	0	0.1	0.4	0.8	0.3	0	0.6	0.5	0.6	0.8	0.2

The results of the maximum accumulated incidence or the optimal way of incidence of the cause a_i to the effect b_j,.

The results are the following: C1-C1-E3-E10, C2-C2-E8-E1, C2-C2-E12-E6, C8-C6-E12-E11, C10-C5-E6-E5, C10-C5-E12-E11.

5 Analysis

The matrix includes the cumulative effects of first and second generation and shows indeed that all variables without exception have a cross effect of, at least, an 80 % multiplier effect. That is, it is generally observed that the variables corresponding to the principles have a direct impact through some element brought about by specific variables to climate change adaptation [31].

The direct implication of these considerations caused the need for decisional models that take into account the entire battery of variables involved in the study but apparently the incidence relation is not appreciated in advance.

This means that prospective studies will be necessary to consider these elements primarily, though apparently not reflect direct effects if they cause very high levels and implications.

The incidences of first generation are very clear on the guiding principles that correspond to the process of adaptation, and the incidences hidden variables override conceptualization of adaptation.

In the case of strategies, we present a direct impact on short-term strategies, this means more tactics that obey monitoring activities and early warning systems, together with those that affect or have higher risks regarding people. The incidences of hidden strategies are primarily those involving technology and infrastructure and risk of damage.

6 Conclusions

The methodology proposed in this study for recovering forgotten effects in the field of adaptation of climate change policies, such as the valuation of actions aimed at its integration within a global legality and socio-political framework, has obtained a wide interest.

We have shown a useful tool for the analysis and evaluation of policies and strategies in various fields, here tested in the field of climate change, specifically in adaptation strategies.

This model specifies the incidents that produce multiplier effects important for decision-makers in the decision-making of high uncertainty.

The necessity of a precise analysis of the exterior position of a cooperating country and the preferences in the policies aimed at its incorporation into a common market is one of the most complex and unavoidable targets.

This model specifies the actions/causes that should be performed and that would result in a higher accumulated incidence to the aspects that are proper for each country in its foreign economic policy.

We emphasize the E6 strategy "Establishing monitoring/early warning systems" and the C8 principle "Modify existing policies, structures and processes", multiplier elements, finding a 90 % multiplier effect and tapping in its class.

The policies need to be directed under the set of regulatory principles. This is of great importance for the evaluation of management activities.

Adaptation should be seen as a process involving scarce resources and political skills [31].

Adaptive elements that are used in strategies obey those involved in the definition, i.e., selection, planning, priorities, and not in the implementation of solutions.

The principles of adaptation support a sustainable development in decision making under any context, providing a solid basis for identifying adaptation priorities and promoting the creation of institutions.

In times of crisis in which the limitations of resources and capacities are high, it is necessary to focus efforts on those strategies and decisions, the positive effects are higher and achieve more reach.

Adaptation measures to climate change should be marked in inclusive and sustainable policies, integrating the transition to a sustainable economy in collective decision processes.

Large cities are a reference to undertake adaptation measures to climate change, exploiting comparative advantages they have, and the potential for innovation, both as victims and as responsible for the climate change.

References

1. R.A. Pielke, J. Gratz, C.W. Landsea, D. Collins, M.A. Suanders, R. Musulin, Nat. Hazards Rev. **9**(1), 29–42 (2008)
2. Munich Re, *Munich Reinsurance Group* (2006)
3. IPCC, *Climate Change 2007* (Cambridge University Press, Cambridge, 2007)
4. P. Vellinga, E. Mills, G. Berz, L.M. Bouwer, S. Huq, L.A. Kozak, J. Palutikof, B. Schanzenbächer, G. Soler (Cambridge University Press, Cambridge, 2001), pp. 417–450
5. E. Mills, *Ceres Report* (October 2007)
6. A.F. Dlugolecki, Thoughts about the impact of climate change on insurance claims, in *Workshop on Climate Change and Disaster Losses*, ed. by P. Höppe, R.A. Pielke Jr. (Hohenkammer, Germany, 2006)
7. B. Van den Hurk, A.K. Tank, G. Lenderink, A. van Ulden, G.J. van Oldenborgh, C. Katsman, H. van den Brink, F. Keller, J. Bessembinder, G. Burgers, G. Komen, W. Hazeleger, S. Drijfhout, *KNMI Scientific Report*, WR (2006-01)
8. W.J.W. Botzen, L. Bouwer, J.C.J.M. van den Bergh, Climate change and increased risk for the insurance sector: a global perspective and an assessment for the Netherlands. *Springer link* 577–598 (2010), http://www.icrea.cat
9. H.C. Kunreuter, E.O. Michel-Kerjan, Paper prepared for the University of Pennsylvania law conference on climate change, 16–17 November 2006
10. L.A. Zadeh, Inform. Control **8**, 338–353 (1965)
11. L. Canós, V. Liern, Eur. J. Oper. Res. **189**, 669–681 (2008)
12. H.J. Chen, S.Y. Huang, C.S. Lin, Expert Syst. Appl. **36**, 7710–7720 (2009)
13. A.M. Gil-Lafuente, *Lógica difusa en el análisis financiero* (Springer, Berlin, 2005)
14. A.M. Gil-Lafuente, J.M. Merigo, *World Scientific* (2010)
15. A. Kaufmann, *Introduction to the Theory of Fuzzy Subsets* (Academic, Nueva York, 1975)
16. A. Kaufmann, J. Gil Aluja, *Introducción a la teoría de los subconjuntos borrosos en la gestión empresarial* (Milladoiro, Santiago de Compostela, 1986)
17. S.T. Li, H.F. Ho, Expert Syst. Appl. **36**, 411–422 (2009)
18. H.C. Lin, F.C. Lin, T.Y. Hsiao Lin, Y.C. Lin, Expert Syst. Appl. **36**, 4535–4540 (2009)
19. J.M. Merigo, A.M. Gil-Lafuente, Fuzzy Econ. Rev. **13**, 17–36 (2008)
20. J.M. Merigo, A.M. Gil-Lafuente, Inform. Sci. **179**, 729–741 (2009)
21. J.M. Merigo, A.M. Gil-Lafuente, Inform. Sci. **180**, 2085–2094 (2010)
22. J.M. Merigo, A.M. Gil-Lafuente, L. Barcellos, Fuzzy Econ. Rev. **15**, 25–42 (2010)
23. A. Kaufmann, J. Gil Aluja, *Modelos para la investigación de efectos olvidados* (Milladoiro, Vigo, 1988)
24. GEF-The Global Environment Facility (2013)
25. A. Kaufmann, J. Gil Aluja, *Técnicas especiales para la gestión de expertos* (Milladoiro, Santiago de Compostela, 1993)
26. A.M. Gil Lafuente, *Nuevas estrategias para el análisis financiero en la incertidumbre* (Ariel, Barcelona, 2001)

27. A. Kaufmann, J. Gil Aluja, *Grafos neuronales para la economía y la gestión empresas* (Pirámide, Madrid, 1995)
28. A.M. Gil Lafuente, L. Barcellos, Cuadernos del CIMBAGE. **12**, 23–52 (2010)
29. A. Prutsch, T. Grothmann, I. Schauser, S. Otto, S. McCallum, Guiding principles for adaptation to climate change in Europe (November 2010)
30. H. McGray, A. Hammil, R. Bradley, *Weathering the Storm: Options for Framing Adaptation and Development* (World Resources Institute, Washington, DC, 2007)
31. B. Lim, E. Spanger-Siegried, *Adaptation Policy Frameworks for Climate Change: Developing Strategies, Policies and Measures* (Cambridge University Press, New York, NY, 2004)

The Challenge of Integration Against the Big International Reconstitutions: Role of Science, Culture and Sport

Mohammed Laichoubi

Abstract The twenty-first century is supposed to be the one for Integration, enabling adaptation and connecting with these multiple new societies, the large human groups.

The mobility of researchers and fluidity of exchanges are also essential vectors in the transfers of technologies, which promote the return of the entrepreneurship...a cultural melting pot leading to a new rise in the splitting of innovation.

In this respect, it is important to emphasize the need for coherence between the models of institutional organisation, the chosen socio-cultural order and the approaches in the field of sciences, culture and performance sports.

Keywords Globalization • Processes of colonization and decolonization • Resurgence of cultures

Globalisation has entered our imaginary and its irreversibility is enshrined in our credos.

This is not so much in its principle that it is rebutted but it is rather in its approaches that it is overturned.

Hence a more informed analysis of the world changes reveals the slow process in the previous century by becoming noticeably more decisive with the decolonisation movements.

132 countries have joined the UNO at its inception. It is the 2/3rd of the world population that freed itself from the colonial empires of which two occupy more than 54 millions of km^2.

M. Laichoubi (✉)
Correspondent Academician of the Royal Academy of Economics and Financial Sciences of Spain
e-mail: mlaichoubi@yahoo.fr

A.M. Gil-Lafuente and C. Zopounidis (eds.), *Decision Making and Knowledge Decision Support Systems*, Lecture Notes in Economics and Mathematical Systems 675, DOI 10.1007/978-3-319-03907-7_9, © Springer International Publishing Switzerland 2015

The Commonwealth organisation previously gathering the territories under the British occupation accounts for 53 countries representing 1.8 billion inhabitants making 30 % of the world population whereas the Portuguese empire gathered 49 states which are currently independent.

The processes of colonisation and decolonisation have led to an intermingling of populations unprecedented in the history of humanity.

The utilisation of labour, from populations under colonial domination in order to build a powerful empire as well as its metropolis and the inclusion by force of men from these dominions and colonies in the armed forces in wars started by these powers, have ended up in important displacements of populations The British Empire in itself accounts for 2.5 millions of people who have served in the British army in 1914.

In addition to these large movements of populations of the colonial and post colonial periods, other waves of immigrants have emerged as a result of the favoured international economic options in the aftermath of the two wars.

For many nation states, the obvious consequences of these movements of population had a resounding impact on the internal socio-cultural order which is constantly shaken and questioned on its conceptual plan failing to take into account the readiness of these populations.

In effect, the rigidity of the philosophies advocated for centuries based on the stratification of cultures alongside the racial divide shaping part of these opinions will not ease up the required transformation of these societies.

Moreover, all the above are to be seen within the constraints linked to the unprivileged social conditions of the new populations coming out of the old empires.

The reviewing of some of the opinions in the metropolis will appear to be destructive when they are confronted to the reality of a social mapping for its most part reconstituted and diversity not expected, with which they will have to cohabit.

Facing this debate on integration in general and the cohabitation of identities, some nations have been better prepared than others especially in terms of their institutions.

Such is the case with the United States of America whose institutional machinery despite some failure related to the racial riots, remains more flexible and efficient in so far as it accommodates within its conceptual choices the integration concerns of all its social groups.

From this point of view a good number of European nation states start off this debate painfully and with difficulty.

In this connection, the General De La Maisonneuve, director of a seminar on Strategy at The University Paris XI reckons that:

"The Nation State is considerably weakened. Being a vertical and authoritative state, it has been conceived for war and built mostly for this purpose; it is particularly badly organised to solve social and economical questions which in our times are exceptionally complex."

However these very national spaces, which have largely impelled this globalisation, find themselves faced with a twofold constraint:

On the one hand, the historical constraint characterised by an internal social reconstitution which has been set up under conditions of inequality largely emphasised during the colonial periods and made possible by the incorporation of indecent cheap labour.

On the other hand, an economical approach though keeping the same options evolves in a different context of an unbridled liberalism which prevailed before the world economical crisis.

This situation has resulted in an internal social order characterised by important fractures and insidiously stimulating a two tier citizenship leading into the disruption of the national cohesion and the inefficiency in the world economics of those countries.

The second constraint is connected to the image sent to the rest of the world which could prove to be ambiguous and paradoxical. That is a space wanting globalisation but not taking on the requirements. A space rather difficult to fit in with the new social mappings which define the twenty-first century and witness the emergence of a certain number of countries including China, Brazil, India, South Africa and so on. These countries will weigh heavily in the shaping of the new world.

These changes see the resurgence of cultures long marginalized and now having to be dealt with.

However, the twenty-first century is supposed to be the one for Integration, for enabling adaptation and connecting with these multiple new societies, the large human groups.

How therefore do we equip ourselves in order to face these upheavals and confront the new challenges? Is it by forging new concepts or by acquiring new means of thought to finally find new approaches? Or are the latter already put into practice elsewhere in other fields.

In the light of the reality that motivate them, the world of sciences, culture and sports seem to be already ahead on these questions.

However, prior to this questioning it is worth stressing that the organisation of the society, its capacity of innovation and research, its performances, are also linked to the socio-cultural model it follows, to the cohesion and the fluidity of its social relations.

From this point of view some European models appear to show some handicaps which do not enable them in some cases to use all their capacities and notably the human resource part of which is marginalised and often unqualified as a result of the social fractures experienced by these communities.

In order to seize the importance of this problem, it is useful to remind ourselves that the field of scientific research and of innovation requires the mobilisation of an important and motivated human potential.

In 2009, China had 1.74 millions researchers and human resources. The USA counted for 1.43 millions whereas the EU had 1.36 millions.

Obviously, prior to evaluating the importance of these figures, we have to leverage them from the base of the pyramid of the communities before reaching the top where the researchers are.

Moreover another important question not always raised in the definition of politics concerns the level of openness to international research that is in it the condition for the performance of the socio-economical models.

The Chinese have delivered the State Awards for scientific progress and international cooperation in sciences and techniques for 2012 to five foreign scientists from USA, Canada, Japan and Denmark.

Fu Xialan, director of the developing Technology and Management Centre at Oxford University stressed that:

"The ancient model according to which the United States, Japan and the EU constituted the main contributors to science is over."

According to the European Business Institute, China occupies the first position among 141 world economies on the basis of the global index of the effectiveness of innovation, an indicator that compares the investments incurred and the yielding of innovation.

In addition, it sets up scientific and technological co operations throughout the world.

Within its Academy of sciences in 2010, out of 694 members 56 are foreigners.

On another issue of interest to our analysis, following the evidence of the failure of internal socio cultural models, we are interested in the approaches advocated at the level of international relations.

We notice that the ultra-liberal economical models proposed in sub-Saharan and Sahelian Africa have collapsed (or led to bankruptcy) in Ivory Coast, Senegal, Mali and Niger etc.

The human tragedies linked to hunger and to illegal immigration make the everyday news on the European shores and the Mediterranean and give a somewhat tarnished image of international relations between Europe and Africa.

This is even more the case because if the USA invests 20 % in their southern neighbours' space and Japan 23 % in its neighbouring states, Europe on the contrary only invests 2 % on its southern shore.

China on the other hand comes in with a vision of a completely renewed cooperation with Africa that is more dynamic and more audacious totally different from what was initially advocated between Europe and Africa.

Indeed in the field of scientific research, it has drawn up the following axes with Africa.

- Reinforcement of the means of subsistence and economical development
- Technical cooperation in water conservation and management, sanitation, genetic improvement of crops, renewable energy.
- More than 100 partnerships of research have been identified.
- Over 100 post doctoral scientists will be involved with research in the Chinese techno parks (22,000 USD will be allocated by means of instruments for research at the end of the study).

The USA with its own model is not lagging behind. Jacques Henri Cost (Research and development dynamics in the USA, origin and evolution of the American system of innovation) reckons that:

"The mobility of researchers and the fluidity of exchanges are also essential vectors in the transfers of technologies which promote the return of the entrepreneurship...a cultural melting pot leads to a new rise in the splitting of innovation . The latter does speed up because the R&D is strongly internationalised, the USA attracting the foreign scientists to their universities..."

In this respect, it is important to emphasize the need for coherence between the models of institutional organisation, the chosen socio-cultural order (diversity or uniformity as well as immigration politics) and the approaches in the field of sciences, culture and performance sports.

Paradoxically, and against the approaches which define the internal order on the socio-cultural plane, Europe is less cautious as far as performance sports, cultural creativity and scientific research are concerned.

In fact in sports, individuals are included, considered, respected according to their talent and competences.

Even in England, despite the Malvinas conflict in the 1980s, nobody would question Messi on his Argentinean origin or even Maradona who was welcomed in Oxford. Zidane has been elevated as an icon in several countries especially in France where he has been elected as the preferred personality of French people despite his origin as the son of an Algerian workman coming from a country that won its independence after a horrid war which lasted 7 years.

In Spain, Costa the Brazilian player of Athletico de Madrid rather than playing for the selection of his own country has chosen the Spanish national team as a sporting challenge. The latter has succeeded in bringing together in a perfect cohesion, Catalans and castellans who display excellent results in the field of sport performance while in the social field divergences separate them.

The examples abound in the sport of performance where approaches are often dynamic, audacious and innovative hence distancing themselves from short sighted visions.

The model is much slower in cultural production sometimes displaying a sense of collective ownership. However the artist becomes quickly the property of a group if not universal and escapes the straitjacket in which we want to confine him. A clear example is Gerard Depardieu, a French actor, who has been a subject of contempt in France as a result of a certain way of apprehending nationalism following a debate on the new taxing system and ending up with the actor being offered other nationalities from Belgium and Russia.

It is obvious that it is from these models of organising performing sports, science and research and development as well as cultural renewal that we have to draw new concepts for the achievement of the twenty-first century's societies.

Societies where mixing and confronting do not require giving up one's identity.

Deletion of identities is not, as some thinking trends suggest, indispensable for the success of globalization.

On the contrary identities enable the dynamics of progress by aggregation.

It is through contact with one another that human beings can remove the barriers and understand each other. It is not by denying the other that they will manage themselves and equally it is through their own cultures that they will shape new compatibilities which will enable them to open up.

It is this capacity of reading and confronting the other which will enable them to identify and to mutually re-invent themselves.

It is this approach that we need to promote in order to create a dynamic that will enable the human being to evolve, to transcend and question oneself freely by interacting with one another hence contribute to the enhancement of humanity and generate new societies by banishing the demeaning categorisations which, whether overtly or insidiously, advocated the favouring of some against others.

Complexity as Interplay Between Science and Sport

Francesco Carlo Morabito

Abstract Complexity is a concept that clearly escapes to a general definition and any universal metric. In this work, it is argued about the relevance of this concept in modern science with particular emphasis to the interplay with sport and some related applications. In particular, it is claimed that suitable measures of complexity can be of help in analyzing and monitoring the performance of athletes as well as to predict possible injuries during exercise. As a matter of fact, some electrophysiological signals that can be easily acquired through noninvasive procedures may contain relevant information about the status of the muscles of athletes both during exercise and at rest. In particular, the analysis of the electromyogram (EMG), whose amplitude has been used for researches on myo-electrically controlled elbows, wrists, and hands, can also yield important clues in biofeedback applications, in ergonomic assessment. For example, in biomechanics, it is used to estimate the torque produced about a joint. Complexity can detect the level of compromising of muscle force production with fatigue. Thus, the extracted information can be useful for predicting muscular injuries and to properly guide athletes' recovery. In this paper, it is highlighted the role of complexity measures carried out on the related physiological time-series.

Keywords Complexity • Alzheimer's Disease • EEG • EMG • Rehabilitation Engineering

F.C. Morabito (✉)
Department of Civil Engineering, Energy, Environment and Materials (DICEAM), University Mediterranean of Reggio Calabria, Reggio Calabria, Italy
e-mail: morabito@unirc.it

A.M. Gil-Lafuente and C. Zopounidis (eds.), *Decision Making and Knowledge Decision Support Systems*, Lecture Notes in Economics and Mathematical Systems 675, DOI 10.1007/978-3-319-03907-7_10, © Springer International Publishing Switzerland 2015

1 Introduction

In the book "Six Memos for the Next Millennium", which collects a series of six lectures that the famous Italian writer Italo Calvino was invited to give at the Harvard University (USA) in the Academic Year 1985–86 (*The Charles Eliot Norton Lectures*), Calvino identified the six qualities that literature should have kept in the millennium which was about to begin. Actually, the sixth was never written on paper, since he died in September 1985. The above mentioned qualities were: Lightness, Quickness, Exactitude, Visibility, Multiplicity and (perhaps) Consistency. In this sort of literary "Manifesto", Calvino reported about those "qualities", each one of them representing a property "winning" on its opposite in the successful literature of the new millennium. Each memo acts as a guideline for life and creativity. Although Calvino's book was primarily written with regards to literature, his book can also well relate to the fields of science, art and design.

In this contribution, it will be discussed about a new, different, quality that, in my opinion, can be of special relevance in modern science and culture, including the themes analyzed in this book, i.e., the relationships among culture, science, and sport.

This quality, whose values are here upheld, is complexity. This is not because its opposite, "simplicity", is to be considered any less compelling or may be not entitled to citizenship in science and culture: on the contrary, the most relevant successes of science came from approximations and reductionism. It is just because the latter decades of the last century, through the work of leading scientists, including Gil-Aluja (through the theory of gradual simultaneity), Prigogine, Mandelbrot, Kaufmann and Zadeh, among others, have revealed how the universe depicted and studied by classical physics, which is closed, deterministic, stationary and self-descriptive, is only an approximate description of the real world. The real world is actually better described by non-linear relationships, self-organization and multiple scale representations that interact on many different levels. In this kind of representation, the linearity of cause-effect relationships, fully described by the impulse response in time domain, by spectral analysis in the Fourier domain, and by the Gaussian distribution in probabilistic terms, it is not sufficient to understand what is happening, and thus to correctly predict the future evolution of the system at hand. The well recognized "lightness" and "simplicity" of the physical models, although elegant, is generally too limiting to be implemented in economy, in social sciences, but also in hard sciences. More generally, it is not enough to understand underlying complex systems (e.g., brain, genetic networks, life).

On the contrary, the recognition of the role of complexity as a value per se, and not as an additional complication to the description of models, allows us to identify the solution that will serve in many scientific problems of present interest, in biology, medicine, economics, and sport.

Complexity manifested itself also in the literature: as a definite example, it is possible to think to the Joyce's *Ulysses*, where it is plastically represented in the correlation tables of chapters with the parts of the human body, the arts and the colors; it is also evidently present in the verbal texture of *Finnegans' Wake* or in the theater of Samuel Beckett or Eugène Ionesco. Complexity is certainly a feature of arts: it has been shown in the artworks of Kandinsky, who deliberately introduced complex patterns in his Composition VII, in the fractal paintings of Jackson Pollock or in the complex texture of Picasso's "Las Meninas" reporting about the Velasquez's paintings at the Museu Picasso, in Barcelona [1].

2 About the Nature and Definition of Complexity

According to an attempted general definition, the complexity of a physical system or a dynamical process reflects its capacity to engage in organized structured interactions. Accordingly, high complexity is achieved in systems that exhibit a mixture of order and disorder (randomness and regularity) thus being able to generate emergent behavior.

As previously noted, a generally accepted definition of complexity does not exist, being it often interpreted in special contexts: similarly, several different methods to measure it have been proposed in the literature. This is understandable because of the lack of a unified framework that cuts across all natural and social sciences; however, different kinds of complexity's measures that apply to specific problem domains have been introduced.

By means of some appropriate measures of complexity, it is possible to analyze the dynamic behavior of a system and to compare different systems to each other by means of a common metric. This is especially meaningful for systems that are monitored longitudinally or in special conditions (e.g., fatigue, disease). Differences in complexity among such systems, or in the same system in different conditions, may reveal features of their organization that promote or alter complexity [2].

Some metrics of complexity have been shown to be useful in exploring brain development during the aging process and pathogenesis [3]: in particular, a high level of complexity, in terms of interplay between segregation and integration of different brain areas [4], it is considered the hallmark of wellbeing and normal functioning [5]. In contrast, a reduction of complexity is often related to pathological conditions: as an example, it is well known that during epileptic seizures there is a measurable reduction of complexity in EEG signals mainly associated to a progressive synchronization of different brain regions (see Fig. 1). The increased correlation of the EEG channels is interpreted as a symptom of disease and produces a complexity reduction as a byproduct.

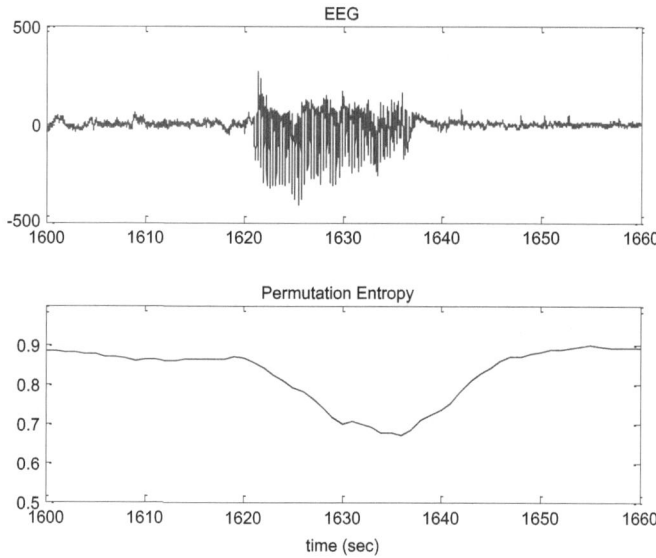

Fig. 1 A segment of EEG signals (electric potential measured by a scalp electrode) showing a typical epileptic seizures (*top*); the corresponding time evolution of the Permutation Entropy, expressed in bits (*bottom*). It is highlighted a loss of complexity in the presence of the synchronization effect generating the seizure. During "normal" state, the complexity is higher than during the critical event

3 Complexity in Science and Medicine

The study and the measurement of complexity is performed through some suitable indicators and markers such as the entropy of the system or of the signal: indeed, entropy is inversely related to the regularity of a signal. In our recent studies [6, 11], we used the concept of *permutation entropy* in order to discriminate among different states of the human brain, during the evolution of some neurodegenerative diseases such as Alzheimer's Disease (AD) [6, 7]. The destruction of neurons and synapses which is determined by the presence of amyloid plaques or tau protein causes a reduction of the connectivity among brain areas and the inability to fully develop the cognitive functions [8, 9]. This neural degeneration can be measured at a macroscopic level by a reduction of the complexity of the EEG signal. Figure 2 shows how the complexity reduction of the EEG signal in AD, with respect to the EEG of a healthy control, implies an enhanced compressibility that, incidentally, may favor the remote monitoring of non-hospitalized patients. In other words, the lower complexity of the EEG signal implies a higher level of redundancy which in turn allows an easy compression and recovery of the signal [10, 11].

The physical and mental well-being of an individual is indeed a sign of complexity: the presence of repetitive or cyclic behaviors is a symptom of malfunctions, and reduces the complexity level. This has been clearly shown in neurological

Fig. 2 Time evolution of a segment of EEG signal for a healthy control (HC) and a patient affected by Alzheimer's Disease (AD). The signal is compressed by an appropriate CS (Compressive Sensing) technique and then reconstructed at a site different from the acquisition electrode. By considering the same Compression Ratio (50 %), it is shown that the reconstruction is more reliable in the AD case, which is a symptom of reduced complexity

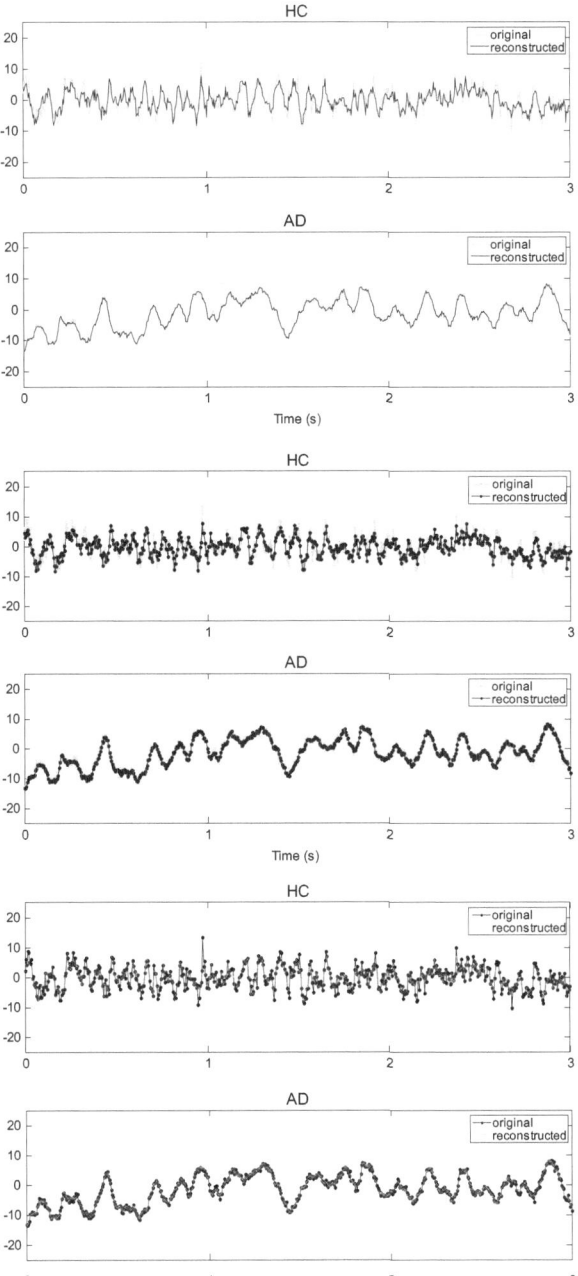

diseases, namely the AD, the Creutzfeldt-Jacob Disease (CJD) and the Parkinson's Disease: in the first case, the disease is characterized by a progressive slowing and loss of complexity [9, 12]; in CJD, the presence of spike-and-wave complexes is highlighted in the complexity metric domain through both a cyclic complexity reduction and a brain lateralization; finally, in the Parkinsonism, the so called tremor manifests itself through an altered pattern of complexity.

4 Complexity in Sport

In this paragraph, various scientific problems related to sport are briefly summarized, aiming to show the potential relevance of the concept of complexity and of its measure in such kind of applications.

The electromyographic (EMG) signal is a de facto standard in order to have access to the physiological processes involved in producing joint movements. The information that can be extracted from the surface EMG signal (sEMG) can be exploited in various sport applications. It is indeed routinely used to control rehabilitation devices or to analyze the biomechanical response of the musculo-skeletal system during the execution of physical exercises through different movements of the arms and the legs. In the latter case, important features about the natural and pathological functioning of the neuromuscular system can be obtained aiming to assess the posture and typical athletes' movements. One of the relevant aspects of the above mentioned biomechanical problems is the detection of the onset of muscle contraction, typically carried out by visual inspection and by implementing a statistical technique based on a suitable thresholding. Because of many reasons, among which abnormal high muscular activity at rest, different kinds of noise and operator fatigue, the procedure is not reliable and the definition of some feature based on complexity measures can be beneficial [13, 14].

In general, dynamic analysis of sEMG is no more limited to the analysis of timing of muscle activation and of the force produced by the muscles being monitored: indeed, the use of markers based on complexity has made it possible to assess muscle fatigue during dynamic contractions. Furthermore, this kind of analysis is useful to detect in advance disorders of neuromuscular transmission, to associate sEMG complexity signatures with myopathies and abnormal muscle activity. It is evident the economic impact of these kinds of approaches: to predict a possible muscular injury has an impact on the management of athletes and possibly on the preservation of the society assets.

These measures are particularly important in the case of sport, where the athlete is subjected to a series of exams aiming to monitor the status of the body and to improve athletic performance in sports competitions.

An example that refers to the sport is the people's postural stability and control, i.e., the set of automatic measurements that the body and the proprioceptive system take simultaneously to determine the position of the body and its balance. The number of elements to be taken into account is very high: a redundant stream of information comes from the sense organs, muscles, and bi-directionally from the brain. This continuous information flow is invariably changing, and the recorded data come from underlying non-stationary processes. The damage of the complex system, at different levels, can be detected by means of complexity measures though the recording of a set of signals that can be then analyzed. The limp generates, for example, a periodic/cyclic gait signal, whose complexity is low because it is highly predictable. In competitive sports, the early detection of a disturbance of these signals can be crucial to avoid a muscle injury, which also has economic consequences.

It has been observed that there is a loss of complexity in the human body as we age. This is more prominent in the muscle strength and activity [15]. Surface Electromyogram (sEMG) reflects the strength of muscle contraction. This age-related changes in sEMG have been associated with a reduction in the number of muscle fibers and a drop in the ratio of type II muscle fibers. Various studies have shown that there are changes in the muscle due to age. In order to verify and identify the reasons for these changes, experiments were conducted on subjects by using the concept of fractal dimension of sEMG, which is indeed a measure of complexity [16]. Results show that there was significant change in the fractal dimension of sEMG and this change was observed in both experimental and simulated sEMG. The same approach can be easily translated to the evolution of muscles in athletes due to different stress conditions or to the use of particular drugs, possibly originating the neuromuscular disorders [17, 18].

In Fig. 3, it is shown that the complexity of a Biceps Brachii (BB) sEMG signal is influenced by fatigue. The Sample Entropy (in bits) of the signal was measured over 50 time scales ($\tau \in (1,50)$) and a piecewise-linear regression has been performed that minimized the squared-error between predicted and observed SE. There is a single breakpoint demarcating two log-linear scaling regions for the analysis of individual time-series (indicated as critical τ, τ_c). It has been shown that muscular fatigue or contraction intensity affected the short-term complexity [19, 20].

Ultimately, the scientific tools that allow us to identify changes in complexity in the typical signals of the human body, e.g., EEG, EMG, acceleration, weight, may suggest optimal strategies for monitoring and controlling the performance of the athletes. Consequently, they also have a decisive effect on the economic aspects related to it, which are of special significance with regard to the activities of the Royal Academy of Financial and Economic Science.

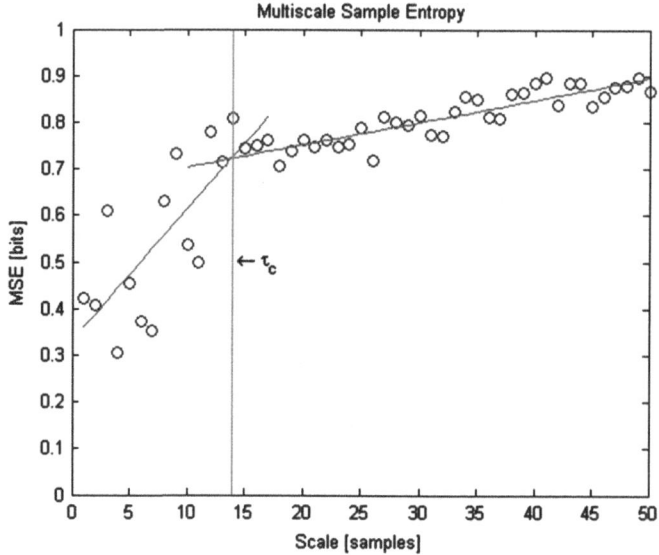

Fig. 3 Dependence on the scale of the complexity, measured by Multiscale Sample Entropy (MSE), for sEMG signals acquired on biceps brachii during activity across fatiguing isometric contraction. The short-term sEMG complexity is reduced because the faster varying control of muscle force production is compromised with fatigue. Reduced entropy is obtained for a less complex signal: healthy biophysical systems are more complex than altered systems. The indicated critical value of scale (τc) discriminates the short-term from the long-term complexity

5   Conclusions

In this article, a summary of the concept of complexity and its application to some aspects of interest for sports has been given. A short presentation of the complexity metrics in different cultural realms (arts, literature, medicine, and biology) has been reported with the aim of convincing about the relevance of this concept and consequently about the interest in automating extracting the related features from any available related readings (i.e., databases, images, time-series). In sports applications, the considered signals often present themselves as typical time evolution of bio-electrical potentials, which are commonly acquired through suitable noninvasive procedures (i.e., EMG). In the case of sport applications, commonly related to biomechanics, surface EMG signals may contain information for a better understanding of strategies underlying human movement and its alterations. It is to be expected that these kind of studies may yield interesting contributions to athletes monitoring and control, for example by interpreting the dynamic amplitude variations of sEMG signals to predict the forces exerted by muscles; this way, they provide an indirect assessment of muscular force and can help to prevent dangerous injuries.

Acknowledgment This contribution represents an extended version of the talk given by the author at the 8th International Meeting of the Royal Academy of Economic and Financial Sciences, held in Barcelona, Spain. The author specially thanks Professor Jaime Gil Aluja for yielding this opportunity.

References

1. F.C. Morabito, M. Cacciola, G. Occhiuto, in *2011 International Joint Conference on Neural Networks*, pp. 2387–2394 (2011)
2. G. Simone, F. Morabito, R. Polikar, J. Appl. Electromagn. Mech. **15**, 291–294 (2002)
3. A.L. Goldberger, L.A.N. Amaral, J.M. Hausdorff, P.C. Ivanov, C.-K. Peng, H.E. Stanley, Proc. Natl. Acad. Sci. U.S.A. **99**, 2466–2472 (2002)
4. G. Tononi, O. Sporns, G.M. Edelman, Proc. Natl. Acad. Sci. U.S.A. **91**, 5033–5037 (1994)
5. L. Lipsitz, A.L. Goldberger, JAMA **267**, 1806–1809 (1992)
6. F.C. Morabito, D. Labate, F. La Foresta, A. Bramanti, G. Morabito, I. Palamara, Entropy **14** (7), 1186–1202 (2012)
7. C. Bandt, B. Pompe, Phys. Rev. Lett. **88**, 174102 (2002)
8. X. Delbeuck, M. Van Der Linden, F. Collette, Neuropsychol. Rev. **13**(1), 79–92 (2003)
9. J. Jeong, Clin. Neurophysiol. **115**, 1490–1505 (2004)
10. J.-H. Park, S. Kim, C.-H. Kim, A. Cichocki, K. Kim, Fractals Interdiscipl. J. Complex Geom. Nat. **15**, 399 (2007)
11. F. Morabito, D. Labate, A. Bramanti, F. La Foresta, G. Morabito, I. Palamara, H. Szu, IEEE Sensor. J. **13**(9), 3255–3262 (2013)
12. D. Labate, F. La Foresta, G. Morabito, I. Palamara, F. Morabito, IEEE Sensor. J. **13**(9), 3284–3292 (2013)
13. E.A. Clancy, N. Hogan, IEEE Trans. Biomed. Eng. **44**, 1024–1028 (1997)
14. D.A. Gabriel, J.R. Basfor, K.-N. An, IEEE Eng. Med. Biol. Mag. **20**, 90–96 (2001)
15. S. Arjunan, K. Wheeler, H. Shimada, D. Kumar, in *Biosignals and Biorobotics Conference (BRC)*, ISSNIP, pp. 1–4 (2013)
16. G. Naik, D. Kumar, S. Arjunan, in *Conference of the IEEE*, pp. 364–367 (2009)
17. E. Park, S.G. Meek, IEEE Trans. Biomed. **42**, 1048–1052 (1995)
18. R. Sun, R. Song, K.-Y. Tong, *IEEE Trans. Neural Syst. Rehabil. Eng.* (2013). doi:10.1109/TNSRE.2013.2290017
19. J.G.A. Cashaback, T. Cluff, J.R. Potvin, J. Electromyogr. Kinesiol. **23**, 78–83 (2013)
20. D. Osipova, J. Ahverinen, O. Jensen, A. Yilikoski, E. Pekkonen, Neuroimage **27**, 835–841 (2005)

Affinity as Basis for Interchangeability Between Athletes

Jaime Gil-Lafuente and Anna Maria Gil-Lafuente

Abstract As consequence of the recessive and depressive process of economic activity, sports entities have been forced to review its management in order to reduce financial costs. This requires a very strict administration from an aspect that generates one of the greater financial needs throughout the exercise: the acquisition of the rights over the athletes. But in addition, the selection of an athlete becomes more complex when it refers to a team sport. In team sports, in addition to having good professionals, groups of interchangeable athletes must be hired, in order to make substitutions due to fatigue, injury or various incidents. From a scientific perspective, algorithms are applied to solve some of the problems when placing financial resources in hiring a team athlete [Pichat (Inform. Process. 69, 1969), Huang et al. (Inform. Process. Lett. 99(4):149–153, 2006)]. In this paper we address the formation of groups of athletes with a high degree of interchangeability when there is inaccuracy in the information available. We will use the concepts of similarity, similitude and affinity. The objective is the adaptation of groups of players reducing financial cost to the team but maintaining the highest reachable performance.

Keywords Affinities • Fuzzy Subset • Fuzzy Logic • Grouping • Sport Management • Uncertainty

J. Gil-Lafuente (✉)
Comercialització i Investigació de Mercats, Universitat de Barcelona, Av. Diagonal, 690, Barcelona 08034, Spain
e-mail: j.gil@ub.edu

A.M. Gil-Lafuente
Faculty of Economics and Business, University of Barcelona, Barcelona, Spain
e-mail: amgil@ub.edu

A.M. Gil-Lafuente and C. Zopounidis (eds.), *Decision Making and Knowledge Decision Support Systems*, Lecture Notes in Economics and Mathematical Systems 675, DOI 10.1007/978-3-319-03907-7_11, © Springer International Publishing Switzerland 2015

1   Previous Aspects to the Proposed Schemes

We start from the simplest notion of similarity to try to reach the deepest affinity. As it is known, even though similarity and affinity express an equivalent idea, there is a difference that we consider essential: the similarity indicates the existence of a certain "resemblance" between two "objects", in our case: athletes. Affinity, however, indicates the good "resemblance" of athletes, 2 on 2, 3 to 3, ..., n to n, in relation to certain characteristics that have been specified (although sometimes inaccurate). The concepts stated illustrate the following:

- For affinity, if two or more athletes may be associated in relation to explicit criteria.
- For similarity, the association only occurs between two athletes.

The similarity is based on the notion of distance [1–4]. The establishment of groups with sufficiently homogeneous elements determines that the smaller is the distance the greater the similarity results. From this statement, it is sought between n different athletes those who are the closest, thus resulting interchangeable in some α level of similarity. The purpose is to obtain subsets; each one made by two athletes, which reach the required level of interchangeability.

The relationships of distance, complementarily of "approach", are commonly represented on a square, symmetric and reflexive matrix, which is considered of similarity. In it, athletes are placed as elements of the rows and columns of the matrix. The degree of similarity between one and the other athletes, two by two, is displayed in the number shown in each of the cells of the matrix.

We have used an endecadaria scale from 0 to 1, so that the closer the valuation "approaches" to the unit, the greater the similarity is. The athletes are always considered in pairs.

We then seek groups that can hold more than two athletes and wherein each group with two or more athletes meets the condition of transitivity. This way we certify that if athlete A gets close or is "comparable" to B, and B is "close" or "comparable" to C, then A can also be "comparable" to C. The word transitivity is particularly important here, because in our case, the coherence of the group of athletes requires to have the properties of symmetry, reflexivity (each athlete has the maximum similarity with itself) and transitivity. These requirements lead us from the notion of similarity to similitude.

Continuing, we obtain these symmetrical, reflexive and transitive sub-relations, from known similarity relations that can be expressed by a matrix, which evidently will be squared, symmetric and reflexive.

To achieve this objective, one can start from the decomposition of the similarity relation into maximum similitude sub-relations using Pichat's algorithm [5, 6]. This algorithm requires that the initial matrix to have the properties of reflexivity and symmetry. Hence the use of similarity relations obtained from the notion of distance.

2   The Transition from Similarities to Affinities

A new algorithm, which passes from the similitude to the affinity, is presented. This algorithm can be applied in any kind of $E_1 \times E_2$ relations, even when the Card $E_1 \neq$ Card E_2. This is the named "algorithm for maximum inverse correspondence" [6], it can be used to obtain affinities in sport's environment. This algorithm is structured by the following phases:

1. "Between the set of athletes and the criteria set, choose the one having the least number of elements".
2. "From this set with fewer elements a "power set" is built, i.e. the set of all its parts".
3. "Every element of "power set" is forced to match the elements of the set that has not been chosen (for having more elements). This is the so-called "link to the right"".
4. "For all non-empty "link to the right" set, the corresponding of the "power set" that has a greater number of elements is chosen".
5. "The relations between the two obtained sets form a Galois lattice [7], which, in addition to present the different homogeneous groups, allows a perfect structure of them".

It has been shown, from the notion of distance, the convenience to find what we call similitude. In our research this approach permits each of the sport groups formed contain those athletes who are the closest between. On the other hand, using the algorithm of affinities, athletes are gathered by previous chosen criteria. These approaches are therefore different. One allows reaching similarity, mathematically speaking, and the other allows affinities in relation to certain criteria.

3   A First Approach to Affinities with No Precise Numbers

From the process of obtaining affinities based on the estimate of the level of qualities, characteristics and singularities possessed by an athlete by a precise number $x \subset [0, 1]$, [6], an extension is presented [8] that incorporates information expressed in confidence intervals $[x_1, x_2] \in [0, 1]$. This allows uncertainty not to be limited to a specific number (with high possibility of error), but consents higher flexibility, even when minor adaptations are needed.

We start describing each of the most representative qualities, characteristics and singularities of the athletes in a fuzzy subset. In our case, we proceed with a generalization for p athletes:

$$p = \{a, b, ..., m\}. \tag{1}$$

and j qualities, characteristics and singularities:

Table 1 Φ-Fuzzy subset

	A	B	...	N
$P =$	$[\mu_a{}^1(p), \mu_a{}^2(p)]$	$[\mu_b{}^1(p), \mu_b{}^2(p)]$	...	$[\mu_n{}^1(p), \mu_n{}^2(p)]$

Table 2 Φ-Fuzzy matrix

		A	B	...	N
$\left[\tilde{R}\right] =$	a	$[\mu_A{}^1(a), \mu_A{}^2(a)]$	$[\mu_B{}^1(a), \mu_B{}^2(a)]$	...	$[\mu_N{}^1(a), \mu_N{}^2(a)]$
	b	$[\mu_A{}^1(b), \mu_A{}^2(b)]$	$[\mu_B{}^1(b), \mu_B{}^2(b)]$	...	$[\mu_N{}^1(b), \mu_N{}^2(b)]$
	c	$[\mu_A{}^1(c), \mu_A{}^2(c)]$	$[\mu_B{}^1(c), \mu_B{}^2(c)]$	...	$[\mu_N{}^1(c), \mu_N{}^2(c)]$
	$\vdots$	$\vdots$	$\vdots$	$\vdots$	$\vdots$
	m	$[\mu_A{}^1(m), \mu_A{}^2(m)]$	$[\mu_B{}^1(m), \mu_B{}^2(m)]$	...	$[\mu_N{}^1(m), \mu_N{}^2(m)]$

$$j = \{A, B, \ldots, N\}. \tag{2}$$

The fuzzy subsets are, here replaced by Φ-fuzzy subsets for the description of an athlete (Table 1).

$$\left[\mu_j^1(p), \mu_j^2(p),\right] \subset [(0,1), (0,1)], \quad j = A, B, \ldots, N. \tag{3}$$

These Φ-fuzzy subsets can be presented in a Φ-fuzzy matrix (Table 2).

This information allows to use the "algorithm for maximum inverse correspondence" previously mentioned in this document. It is then required to convert the Φ-fuzzy matrix $\left[\tilde{R}\right]$ into one or more Boolean matrices, setting a threshold to be appointed by θ.

Three situations can then occur:

1. The inferior end of the interval is bigger or equal than the threshold:

$$\mu_j^1(p) \geq \theta. \tag{4}$$

This is the case in which, regardless of the real level possessed by the quality, characteristic or singularity within or above the interval, it will always be equal to or greater than the threshold.

2. The threshold is found between the two ends of the interval:

$$\mu_j^1(p) \leq \theta \leq \mu_j^2(p). \tag{5}$$

In this case it is commonly admitted that an athlete possesses the required level of a specific quality, characteristic or singularity if the threshold is not greater than the superior end of the interval. Taking this position is acceptable, yet questionable. For this reason (in certain cases), the criterion of considering the possession of the characteristic by an athlete when the threshold is equal or greater than the mean of the ends of the intervals is adopted.

Table 3 Conversion of the Φ-fuzzy matrix $\left[\tilde{R}\right]$ into a Boolean matrix [B]

	A	B	...	N
[B] =	$B_A(a)$	$B_B(a)$	...	$B_N(a)$
	$B_A(b)$	$B_B(b)$	...	$B_N(b)$
	$B_A(c)$	$B_B(c)$	...	$B_N(c)$
	⋮	⋮	⋮	⋮
	$B_A(m)$	$B_B(m)$	...	$B_N(m)$

3. The superior end of the interval is below the threshold:

$$\mu_j^2(p) < \theta. \tag{6}$$

In this case it is customary to admit that the athlete does not possess the required level of the characteristic, attribute or singularity.

If one of these scenarios is accepted (or any of the commonly used), the conversion of the Φ-fuzzy matrix $\left[\tilde{R}\right]$ into a Boolean matrix [B] immediately occurs (Table 3).

If

$$
\begin{aligned}
B_j(p) &\in \{0,1\} \\
j &= \{A, B, \ldots, N\} \\
p &= \{a, b, \ldots, m\}.
\end{aligned}
\tag{7}
$$

Where Bj (p) is the result of assigning 1 or 0 whether the athlete p possesses the quality, characteristic or singularity j.

A threshold considering different levels for each quality, characteristic and singularity can be incorporated. When this happens we must introduce j thresholds:

$$\theta_j = \{\theta_A, \theta_B, \ldots, \theta_N\}/ \quad 0 \le \theta_j \le 1. \tag{8}$$

The transition to a Boolean matrix from a Φ-fuzzy matrix is subject to certain criteria. The interpretation of these criteria and their utilization is presented below with an example [8].

4 The Hypothesis of the Inferior End Being Greater or Equal to the Threshold

Supposing the existence of a small set of four athletes: $E_1 = \{a, b, c, d\}$. And six criteria to establish the affinities: $E_2 = \{A, B, C, D, E, F\}$.

The Φ-fuzzy subsets that define these four athletes have been previously determined as follows (Table 4):

Table 4 Φ-Fuzzy subsets that define these four athletes

	A	B	C	D	E	F
$a =$	.8	[.3, .5]	[0, .2]	[.6, .9]	.8	[.3, .5]
$b =$	.2	[.5, .6]	[.5, .8]	1	.6	[.9, 1]
$c =$	[.3, .5]	.5	[.7,.9]	[.8, 1]	[.5, .6]	.9
$d =$	0	[.3, .5]	[.6, .8]	.7	[.3, .7]	[.8, 1]

Table 5 Φ-Fuzzy $\left[\tilde{R}\right]$ matrix results

$\left[\tilde{R}\right] =$

	A	B	C	D	E	F
a	.8	[.3, .5]	[0, .2]	[.6, .9]	.8	[.3, .5]
b	.2	[.5, .6]	[.5, .8]	1	.6	[.9, 1]
c	[.3, .5]	.5	[.7, .9]	[.8, 1]	[.5, .6]	.9
d	0	[.3, .5]	[.6, .8]	.7	[.3, .7]	[.8, 1]

Based on this information, the corresponding Φ-fuzzy $\left[\tilde{R}\right]$ matrix results (Table 5):

In order to achieve further simplification, it is considered that the required threshold is the same for all the qualities, characteristics or singularities: $\theta = 0.7$.

The transition from the Φ-fuzzy matrix $\left[\tilde{R}\right]$ to the Boolean matrix [B] is now required. In order to achieve so, it is needed to adopt each of the shown criteria by comparing the θ threshold and confidence intervals.

As a first approach we will accept, that an athlete has the quality, characteristic or singularity when the inferior end of the interval is greater than or equal to the threshold. Therefore, the Φ-fuzzy matrix $\left[\tilde{R}\right]$ allows the creation of a Boolean matrix $[B_1]$ (Table 6).

We then apply the "algorithm for maximum inverse correspondence", obtaining:

$*a \rightarrow AE$	$bd \rightarrow DF$
$b \rightarrow DF$	$cd \rightarrow DF$
$*c \rightarrow CDF$	$\varnothing \rightarrow E_2$
$d \rightarrow DF$	$abc \rightarrow \varnothing$
$ab \rightarrow \varnothing$	$abd \rightarrow \varnothing$
$ac \rightarrow \varnothing$	$acd \rightarrow \varnothing$
$ad \rightarrow \varnothing$	$*bcd \rightarrow DF$
$bc \rightarrow DF$	$abcd \rightarrow \varnothing$

Where
$\varnothing \rightarrow E_2$
$a \rightarrow AE$
$c \rightarrow CDF$
$bcd \rightarrow DF$
$E_1 \rightarrow \varnothing E_2$

Table 6 Boolean matrix $[B_1]$

		A	B	C	D	E	F
	a	1				1	
$[B_1] =$	b				1		1
	c			1	1		1
	d				1		1

Fig. 1 Galois lattice that matches athletes with their qualities

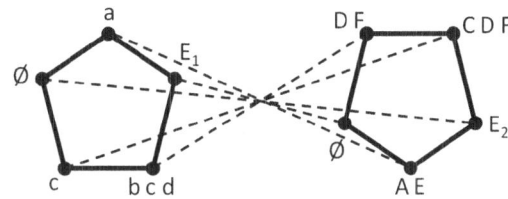

Fig. 2 If the corners of each lattices overlap, we then find the pursued Galois lattice

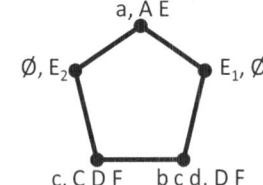

From these correspondences we can find the representative Galois lattice. We present the lattice that matches athletes with their qualities (Fig. 1).

If the corners of each lattices overlap, we then find the pursued Galois lattice (Fig. 2).

5 The Hypothesis of the Mean of the Interval Equal or Greater than the Threshold

We now assume the case in which the level of possession of an athlete's quality, characteristic or singularity is accepted in the middle point of the correspondent interval when:

$$\theta \leq \frac{\mu_j^1(p) + \mu_j^2(p)}{2}. \tag{9}$$

In that case we obtain the following Boolean matrix $[B_2]$ (Table 7).

As is presented in the first hypothesis, we will then use the "algorithm for maximum inverse correspondence" to achieve sufficient basis for valid comparison. We obtain:

Table 7 Boolean matrix $[B_2]$

		A	B	C	D	E	F
	a	1			1	1	
$[B_2] =$	b				1		1
	c			1	1		1
	d			1	1		1

$*a \rightarrow ADE$	$bd \rightarrow DF$
$b \rightarrow DF$	$cd \rightarrow DF$
$*c \rightarrow CDF$	$*cd \rightarrow CDF$
$d \rightarrow CDF$	$abc \rightarrow D$
$ab \rightarrow D$	$abd \rightarrow D$
$ac \rightarrow D$	$acd \rightarrow D$
$ad \rightarrow D$	$*bcd \rightarrow DF$
$bc \rightarrow DF$	$abcd \rightarrow D$

The correspondence in this case is:

$\emptyset \rightarrow E_2$

$a \rightarrow ADE$

$cd \rightarrow CDF$

$bcd \rightarrow DF$

$abcd \rightarrow D$

The lattices corresponding each of the sets and their correspondence are represented in (Fig. 3):

Overlapping the lattices, we then obtain the pursed Galois lattice (Fig. 4).

6 The Hypothesis of Any Point of the Interval Equal or Greater than the Threshold

This is the most commonly accepted case due to its correspondence to the economic idea of interval. In this scenario any point of the interval is accepted as positive due to the uncertainty that is manifested by the absence of any information from the two ends thereof [8].

The Boolean matrix obtained is represented (Table 8).

The next correspondences are found:

$*a \rightarrow ADE$	$bd \rightarrow CDF$
$b \rightarrow CDF$	$cd \rightarrow CDF$
$c \rightarrow CDF$	$abc \rightarrow D$
$*d \rightarrow CDEF$	$abd \rightarrow D$
$ab \rightarrow D$	$acd \rightarrow D$
$ac \rightarrow D$	$*bcd \rightarrow CDF$
$*ad \rightarrow DE$	$*abcd \rightarrow D$
$bc \rightarrow CDF$	

Fig. 3 Lattices
corresponding each of the
sets and their
correspondence

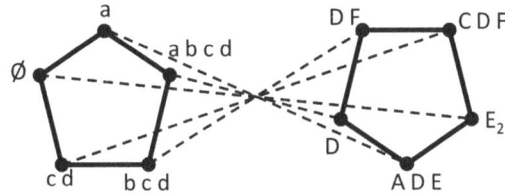

Fig. 4 Galois lattices

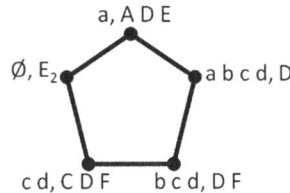

Table 8 Boolean matrix $[B_3]$

		A	B	C	D	E	F
$[B_3] =$	a	1			1	1	
	b			1	1		1
	c			1	1		1
	d			1	1	1	1

We then have:

$a \rightarrow ADE$

$d \rightarrow CDEF$

$ad \rightarrow DE$

$bcd \rightarrow CDF$

$abcd \rightarrow D$

The results obtained at a significance level $\theta = 0.7$ can be presented as reticular as shown in the previous two scenarios. This approach displays different groups of athletes according to their qualities, and also athletes who possess certain qualities at the required or higher level. To the left in the figure the corresponding lattice of the required qualities can be found, to the right the athlete's lattice is shown. The vertices from each lattice are overlapped corresponding to the content of the matrixes (Fig. 5).

Then one of the lattices is inverted and overlapped, turning it upside down matching the other one. The next Galois lattice is obtained (Fig. 6):

In all cases as we move from left to right, athletes groups increase, but the qualities that each group possess decrease. Only in one scheme all the possibilities of joining affine athletes is available, that is in fact an essential condition to choose the one group owning the most important or indispensable criteria considered.

It is easy to observe that according to this last scenario (being the possessed qualities of an athlete expressed as confidence intervals), some athletes who have been admitted the possession of the required level of 0.7 would not reach that

Fig. 5 Reticular shown in
the previous two scenarios

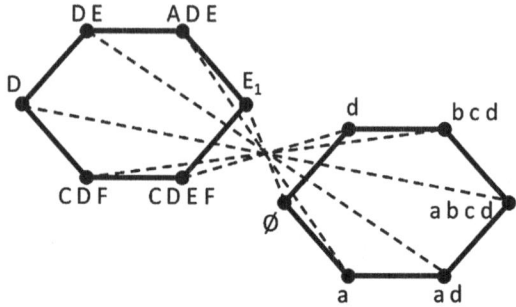

Fig. 6 The next Galois
lattice

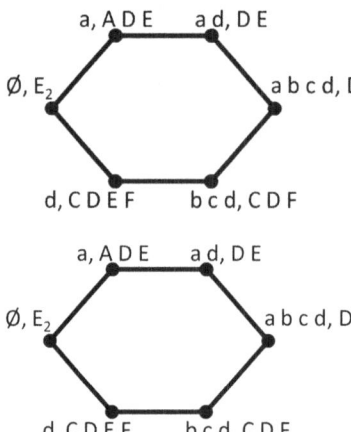

number if the real possession were in the superior or inferior end. It is the case
e.g. of quality C from the athlete B. The estimation has corresponded to the interval
[0.5, 0.8]. In the first scenario the interval must have been [0.7, 0.8] and in the
second one [0.6, 0.8].

7 Conclusions

The proposal presented in this paper aims to address the important problem stated at
the beginning of this study by establishing confidence intervals estimates rather
than precise numbers, assuming the complexity that this generalization suppose.
The scheme with precise numbers is then a particular case of the ones that are
proposed in this work.

The formal aspects that construct the general scheme have been established
based on the description of each athlete by a Φ-fuzzy subset. That information
allows the development of a Φ-fuzzy matrix. The transition from a Φ-fuzzy matrix
into a Boolean matrix had represented a real challenge to those who had used these

elements in high sporting competition. The presented three hypotheses allow covering a wide range of the sports activity of teams, especially when they try to acquire the services of an athlete with the capacity of substitution and interchangeability between other athletes covering a particular position in a team.

With the highly visual schemes proposed, we present a high response "state of the situation" to the decision maker. These schemes allow quick responses, as demanded by the need of substituting an athlete for another one in the middle or throughout a sports match.

The present research shows the possibility of generalizing a methodology and the corresponding algorithms to obtain general affinities and the substitutability of athletes using confidence intervals estimates rather than precise numbers. Changing from precise numbers to confidence intervals has become progressively useful as uncertainty around the topic increases; this uncertainty frequently obstructs accurate estimations on future effects. This contribution is an open invitation to the use of other uncertain, but bounded numbers, such as triplets' confidence, triangular fuzzy numbers, trapezoidal fuzzy numbers, and fuzzy numbers in general, among others.

Substituting precise numbers for confidence intervals fosters the problem of establishing criteria. It has been shown that the establishment of hypotheses makes it possible to move from a Φ-fuzzy matrix to a Boolean matrix, allowing the application of existing algorithms, in our case the "algorithm for maximum inverse correspondence".

Three hypotheses with the most common positions have been presented: optimism, "neutrality" and pessimism.

1. The most pessimistic hypothesis matches the proposal in which every interval must be superior to the threshold; therefore, if the interval's estimation is correct, the threshold will be always covered.
2. The intermediate, middle or "neutral" hypothesis has been made to match the midpoint of the interval. This means (if this is true), there are closer positions to the inferior end that are also considered representative of possessing the qualities, characteristics, or singularities even when they are below the threshold.
3. The optimistic hypothesis (curiously the most used so far, even by ourselves) accepts all space intervals as valid, although reality would place the threshold near or coinciding with the superior end. Implicitly, it assumes the threshold would be set at the inferior end.

As one approach moves from the hypothesis a) to b) and then to c), groups comprise more athletes and/or more qualities. It has been shown that in the transition from the hypothesis a) to b) the group of athletes b, c, d with DF qualities maintains, but a new group of four athletes is created if only one quality, characteristic or singularity D is required. It also adds a new quality D to the athlete a, which in the first hypothesis (more restrictive) did not possessed.

The transition of the hypothesis b) to c) leads all athletes to maintain quality D. It increases the qualities, characteristics or singularities of group b, c, d with the addition of quality C. A new group of two athletes a, d with the qualities, characteristics or singularities D, E is formed. Also the possession of C, D, E, F qualities by the athlete d is shown, however, the qualities C, D, F formerly manifested in c go to the new group of the athlete c with the athletes b and d.

To conclude, we have presented a scheme that generalizes the existing, providing a presumed valid enough solution to the problem of grouping affine athletes due to certain specific qualities.

References

1. J. Gil-Aluja, Fuzzy Set. Syst. **84**(2), 187–197 (1996)
2. A. Popa, J.J. McDowell, Behav. Process. **84**(1), 428–434 (2010)
3. W. Huang, Y. Shi, S. Zhang, Y. Zhu, Inform. Process. Lett. **99**(4), 149–153 (2006)
4. C.W. Duin, A. Volgenant, Eur. J. Oper. Res. **170**(3), 887–899 (2006)
5. E. Pichat, [1969] *Algorithms for finding the maximal elements* of a *finite universal algebra*, in *Information* Processing 68 (Proc. IFIP Congress, Edinburgh, 1968), Vol I: Mathematics, Software, pp. 214–218, North-Holland, Amsterdam
6. J. Gil Lafuente, *Algoritmos para la excelencia: claves para el éxito en la gestión deportiva*, (Milladoiro, España, 2002), pp. 7–59, 221–270
7. L.E. Dickson, Linear groups - With an exposition of the Galois field theory (Courier Dover Publications, 2003), pp. 75–88
8. J. Gil-Aluja, A.M. Gil-Lafuente, Optimal strategies in sports economics and management. (Springer, Berlin, 2010), pp. 1–14

The Efficiency Analysis of Automated Lines of Companies Based on DEA Method

Igor Kovalev, Pavel Zelenkov, and Sergey Ognerubov

Abstract The approach to analyze the efficiency of the technical systems by DEA method is considered. DEA method and its modification are described. The modification allows analyzing the plants by bad outputs. The recycling factories are researched. Finding results are analyzed.

Keywords DEA • Efficiency • Recycling • Analysis

1 Introduction

In recent years an estimation problem of efficient operation of business arises in spheres of product manufacture and production distribution. Frequently appears the problem of comparison and ranking organization departments and firms or organizations on the whole by some characteristics that can't be directly measured [1]. And a general idea of demonstration the degree of analyzing latent quality is formed as a result of particular measured characteristics this quality depends on. Beyond all question efficiency is the main concept. "Efficiency is a most general property of any purposeful activity that from the cognitive point of view uncovers itself by way of the target category and is objectively expressed by the degree of a goal achievement taking into account time and resources expenses." That's why the efficiency estimation of business and organizations operation is very important for making right management decisions.

The efficiency of the garbage recycling companies is analyzed in this paper. Three companies are considered: "Novosibirskiy musoropererabativaushiy zavod"

I. Kovalev (✉) • P. Zelenkov
University Administration, Siberian State Aerospace University, Office A-406
31 Krasnoyarskiy Rabochiy Av, Krasnoyarsk 660014, Russia
e-mail: ogss@rambler.ru

S. Ognerubov
Informatics, Siberian Federal University, 79 Svobodny Prospect, Krasnoyarsk 660041, Russia

A.M. Gil-Lafuente and C. Zopounidis (eds.), *Decision Making and Knowledge Decision Support Systems*, Lecture Notes in Economics and Mathematical Systems 675, DOI 10.1007/978-3-319-03907-7_12, © Springer International Publishing Switzerland 2015

(Novosibirsk city), "Kazanskiy ecologocheskiy complex" (Kazan city), "Stroy plus" (Krasnoyarsk city) [2].

A Swedish equipment line "Presona" (some assembly lines (conveyors), two presses and a plastic bottles recycling device) for garbage sorting and pressing is installed at the plant in Novosibirsk. A sorting technology allows to significantly reduce the amount of solid waste. For example, the plant produces 2 m^3 of waste after processing of 10 m^3 of garbage. The plant in Krasnoyarsk can sort up to 600 m^3 of garbage a day. 240 m^3 of waste remain after the sorting and the separation of useful materials. This waste usually has a bio-decaying nature and much safer than unprocessed garbage.

The plant in Kazan has a garbage sorting installment which was built using Imabe Iberica company's technologies. The sorting process reduces the volume of garbage by 50 %. The pressing process adds an additional 25–30 % reducing.

DEA-method and its modification and realization for an analysis of the technological and organizing aspects of presented companies are considered in this paper. The feature of the realization of DEA-method is that the estimation of the functioning efficiency of companies task is divided by three subtasks. The first subtask is a problem of the company's efficiency evaluating, which provides the maximum output of useful products and materials. The second subtask is a problem of the company's efficiency evaluating, using the sum of the amounts of useful materials after the recycling process. The third subtask is a problem of the company's efficiency evaluating, which provides the minimum output of waste materials after the recycling process. The DEA method modification is used to solve the latter problem.

The efficiency criterion is producing more materials using less input resources. But according to saving environment paradigm, output waste such as air pollution or hazardous waste are being recognized as more and more dangerous and therefore unwanted. This unwanted output should be considered while company's efficiency estimating. Thus companies with a better good (desirable) output and a lesser bad (unwanted) output should be recognized as effective.

2 DEA-Method Usage

The features of garbage recycling companies can be considered and estimated using DEA methodology [3]. These features are [4, 5]:

- The input parameters—raw materials.
- The output parameters—recycled useful product, waste, ecological influence.
- The algorithms for the technological processes can be created using different approximations such as a system which interacts with an environment and factors and many more.
- The multicomponent mass is usually used as a raw material.

- As a result of the automated sorting the components, which cannot be used, remain.
- The efficiency of plant is evaluated using some criterions; therefore it fits DEA method and its lag's bound modification.

3 DEA-Method

The given method estimates the production function that in reality is unknown. DEA-method is based on constructing the efficiency frontier that is an analogy of the production function in the case when the output is not a scalar but a vector, i.e. when there exist some kinds of output. This frontier has a form of a convex hall or a convex con in the space of input and output variables describing each DMU in the system. As follows from method's name, the efficiency frontier envelopes or covers over a scatter of points in the multidimensional space. The efficiency frontier is used as a standard (or a peer) to obtain numerical value of efficiency measure of each object in the researching set. The efficiency degree for the i-th DMU is determined with degree of proximity to the efficiency frontier in the multidimensional space of inputs and outputs. The way of constructing an efficiency frontier is to solve the linear programming problem N times.

Let's describe the main idea of DEA-method on the example. A firm uses two inputs (x_1 and x_2) to produce a single output (y) [6].

If we assume constant returns to scale, we can represent the technology by a unit production possibility curve on a two-dimensional plot. The axes reflect costs for a unit, i.e. the volume of resources x_1 and x_2 per a unit of the output. Consequently, we have a unit isoquantum.

Let the given firm uses quantities of inputs, corresponding to the point P, to produce a unit of the output. Then the technical inefficiency of that firm could be represented by the segment QP. A point Q is a point P projection on the efficiency frontier. At the same time the projection is made towards the coordinate origin. The segment QP is a number by which all inputs (x_1 and x_2) could be proportionally reduced without a reduction in the output (y). This approach of the efficiency measurement is called input-oriented.

The technical efficiency (TE) of the firm is measured as follows:

$$TE_P = \frac{0Q}{0P}. \tag{1}$$

In Fig. 1 A, B, C and D are efficiency points. They form the efficiency frontier. A point P is not technically efficient because it doesn't lie on the efficient isoquantum.

If an object A can produce the definite volume of output from the definite volume of resources, an object B can also produce the same volume of output from the same volume of resources. That's why it is right to project the points corresponding to inefficient objects on the efficiency frontier.

Fig. 1 Production
technology with two inputs
and a single output

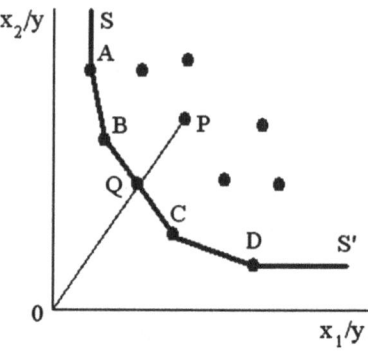

Figure 1 shows that the value of technical efficiency can't exceed 1. Projecting the inefficient object on the efficiency frontier, a hypothetic target object is formed. It is to be efficient. In mathematical sense this hypothetic target object corresponds to a linear combination of the real efficient objects (in this case a real object is a point in multidimensional space). The number of objects included by this combination depends on a number of factors like the input and output variables describing the objects, and values of these variables. The input and output variables of the target object are targets for the inefficient objects.

Let's describe an idea of DEA-method using an example. Assume there exist data for K inputs and M outputs for each of N objects or DMU's (they could be firms, banks, universities). For the i-th DMU they are represented by vectors x_i and y_i, respectively. Then matrix X of $K \times N$ dimension and matrix Y of $M \times N$ dimension are matrixes of input and output parameters for all the i-th DMU. The model is formulated as a linear programming problem:

$$
\begin{aligned}
&\min_{\theta,\lambda}(\theta), \\
&-y_i + Y\lambda \geq 0, \\
&\theta x_i - X\lambda \geq 0, \\
&\lambda \geq 0 \ .
\end{aligned}
\tag{2}
$$

Where θ is a scalar and λ is a $N \times 1$ a vector of constants. The obtained value θ will be the efficiency score for the i-th DMU. It will satisfy $\theta \leq 1$. The same linear programming problem must be solved N times, once for each DMU in the sample.

The presented model (1) is constructed under assumption of constant returns to scale and after its N times solving the efficiency frontier is formed as a convex cone.

Those objects, which have θ equal 1 are located near the bound of efficiency.

4   DEA Method Modification

Consider the model based on DEA method, which is built assuming constant returns to scale [7].

The efficiency bound passes through points which correspond to efficient objects. Let's build the bound but on the contrary. For that, the points which produce the minimum output using the maximum inputs should be found. The new bound will pass through the points which correspond to unprofitable (inefficient) objects. Therefore those objects should be considered as inefficient.

The DEA model for the inefficient bound, which is built assuming constant returns to scale:

$$\begin{aligned}
&\min_{\theta,\lambda}(\theta),\\
&\theta y_i - Y\lambda \geq 0\\
&-x_i - X\lambda \geq 0\\
&\lambda \geq 0 \ .
\end{aligned} \tag{3}$$

Where θ is the lag measurement for an object. θ equal 1 means that object is a part of the bound and therefore is inefficient. θ varies from 0 to 1.

Using this bound we can evaluate inefficiency degree for all considered objects and find the most inefficient of them.

Figure 2a shows the efficiency bound and Fig. 2b shows the inefficiency bound.

The modification of DEA-method can be used to estimate the amount of waste for the recycling plants. It is important to find plants which produce the minimum amount of waste using the maximum amount of input (unsorted garbage).

Here are the criterions which can be obtained using DEA method and the modification to considered sample.

1. θeff is an efficiency measurement for every output for the automated recycling plants.
2. θseff is an efficiency measurement for the sum of the outputs for the automated recycling plants.
3. θwaste.is a coefficient which shows how less waste the plants produce.

Figure 3 shows schematics with inputs and outputs which are used for DEA based analysis. The inputs and outputs are marked as:

- $x1$—the amount(volume) of unsorted raw materials(garbage);
- $y1,\ldots,y7$—the amount(volume) of the sorted, useful materials (paper, carton, glass, plastic bottles, polyethylene, aluminum cans, tin cans);
- $y8$—the sum of all outputs;
- $y9$—the amount(volume) of waste which cannot be used using given technology and installments.

According to given structure of the subtasks, here are their definitions.

Subtask 1. The efficiency of a plant evaluation for every output (glass, paper, carton and so on). It is shown on Fig. 3a.

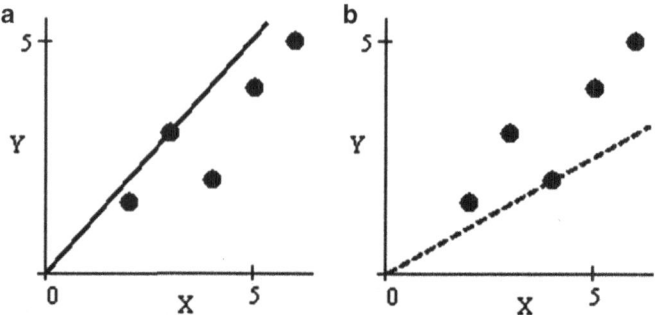

Fig. 2 (**a**) The efficiency bound, (**b**) the inefficiency bound

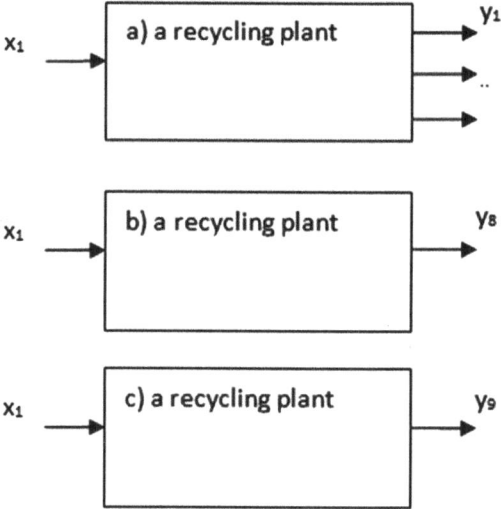

Fig. 3 Schematics with inputs and outputs of the recycling plants

Subtask 2. The efficiency of a plant evaluation for the sum of all outputs. It is shown on Fig. 3b.

Subtask 3. The minimum unwanted waste output evaluation. It is shown on Fig. 3c.

DEA method and the modification are used for schematics which are presented on Fig. 3 as 3a and 3b. According to the methodology, the data sample of several objects (in our case, the recycling plants) is used and then the sets of their inputs ($x1$) and outputs ($y1 \ldots y9$) are analyzed. After that, the efficient objects are defined ($\theta = 1$) and then the efficiency frontier is built. Next, this frontier can be used as a standard for other objects. Using this standard, it is possible to provide the inefficient objects with the parameter correction recommendations.

Using model (3) output $y9$ we can obtain the inefficiency bound.

Table 1 Input, output and efficiency values for subtask 1

Id of a plant	Input Vm³, cubic meters per year	Output in tons							Efficiency measurement θ_{eff}
		Paper	Carton	Glass	Plastic bottles	Polyethylene	Tin cans	Aluminum cans	
1	149,400	9,960	12,450	12,450	7,470	2,490	2,490	249	0.44
2	124,500	15,017	29,238	12,816	19,861	6,724	8,722	840	1
3	60,310	9,660	19,920	16,814	6,105	2,044	3,499	651	1

5 Calculation Results

5.1 Subtask 1

Table 1 shows the data for three plants. They are the plants in Krasnoyarsk, Novosibirsk and Kazan and they are marked as 1, 2 and 3 accordingly.

Plants 2 and 3 are efficient according to the method ($\theta = 1$). Plant 1 is inefficient ($\theta < 1$). It is impossible to show a seven parameter function as a two dimension graph. So let's show ratio of input to each output separately.

Since plants 2 and 3 are efficient, they will be a standard for plant 1. Figure 4 shows that plant 1 does not have any minimum value, only maximum values. It is possible to use the efficiency frontier to get recommendations how to increase efficiency.

5.2 Subtask 2

According to Table 2, the efficiency bound can be built using plant 3 because its $\theta = 1$. It has the best ratio of the summary useful output to the summary input. Table 2 shows that plant 2 has a worse result than plant 3. It should be noticed that plant 1 is more competitive using the summary output than any single output.

5.3 Subtask 3

Table 3 shows that plant 1 is the most environment friendly. Plants 2 and 3 produce more waste after the recycling process and shows less ecological efficiency.

Fig. 4 The ratio of input to output: (**a**) input/paper, (**b**) input/carton, (**c**) input/glass, (**d**) input/plastic bottles, (**e**) input/polyethylene (**f**) input/tin cans, (**g**) input/aluminum cans

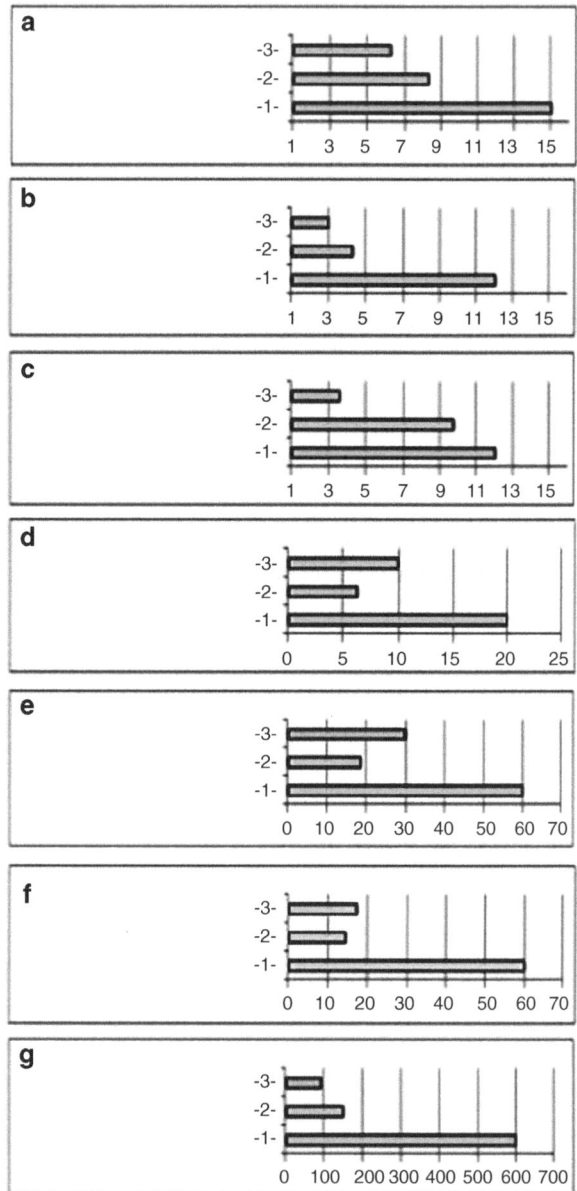

6 Conclusion

The subtasks based on DEA-method were given which provide opportunity to analyze the garbage recycling plants using three criterions. The solution of first subtask provides detailed efficiency analysis. The solution of second subtask

Table 2 Input, output and efficiency values for subtask 2

Id of a plant	Input Vm^3, cubic meters per year	Output Summary output in tons per year	Efficiency measurement θ_{seff}
1	149,400	76,613	0.499
2	124,500	111,718	0.785
3	60,310	68,953	1

Table 3 Input, output and ecological efficiency for subtask 3

Id for a plant	Input Vm^3, cubic meters per year	Output Waste in tons per year	Efficiency measurement θ_{waste}
1	149,400	35,019	1
2	124,500	30,000	0.973
3	60,310	18,000	0.785

provides summary efficiency analysis. The solution of third subtask allows researchers to take a look from a different point of view and to find the negative effects using the waste output. The obtained data allows finding positive and negative points of companies and therefore can be used as additional information in the decision support systems.

References

1. M. Karaseva, A. Novogilov, T. Rukovicina, Vestnik Sibirskogo gosudarstvennogo aerokosmicheskogo universiteta im. akademika M.F. Reshetneva **4**, 40–42 (2011)
2. I. Kovalev, A. Novogilov, T. Rukovicina, Economica i manegement system upravlenia **1**, 36–42 (2011)
3. W.W. Cooper, K. Tone, *Data envelopment analysis: A comprehensive text with models, applications, references and DEA-solver software*, 2nd edn. (Springer, New York, 2007)
4. I. Kovalev, A. Novogilov, T. Rukovicina, Systemi upravlenia i informacionnie technologii **4**, 36–39 (2010)
5. A. Novogilov, Informatica i systemi upravlenia **1**, 98–103 (2010)
6. T.J. Coelli, D.S. Prasada Rao, C.J. O'Donnell, G.E. Battese, *An introduction to efficiency and productivity analysis*, 2nd edn. (Springer, New York, 2005), p. 366
7. T. Rukovicina, Vestnik Sibirskogo gosudarstvennogo aerokosmicheskogo universiteta im. akademika M.F. Reshetneva **24**, 74–77 (2009)

The Expertise on the Valuation Process Applied to the Discounted Cash Flows

Anna Maria Gil-Lafuente and César Castillo-López

Abstract The aim of this paper is to contribute to the analysis on how to measure the value of a company through different techniques which have been used in the literature. We propose a methodology based on the expertons and fuzzy logic. We have investigated different applications of fuzzy methodology, with the aim of deciphering what could work best to determinate the value of a company in the process of valuation. Following the theory of the expertons, we have used the discounted cash flow method to assess organizations. The theory of the expertons is considered, according to the studied literature, an extension of the concept of probabilistic set for uncertain environments, which cannot be measured with exact numbers, but could be measured with numeric intervals. The conclusions in the paper will allow future academic researchers one step forward in the eradication of uncertainty on traditional methods of business valuation, through the application of the fuzzy logic algorithms presented in this article, allowing us to move one step beyond in the valuation of companies.

Keywords Valuation of companies • Uncertainty estimation • Subjective variables • Probabilistic sets

1 Introduction

The main problem identified in the literature on business valuations, lies with the subjective factors involved in the determination of the value of the company. It could deepen to try to explain not only the financial data involved in the final value, but also experimental evidence produced by speculative risks [1]. Since the beginning of the discipline it has been great effort to differentiate the concepts of price

A.M. Gil-Lafuente (✉) • C. Castillo-López
Faculty of Economics and Business, University of Barcelona, Barcelona, Spain
e-mail: amgil@ub.edu

A.M. Gil-Lafuente and C. Zopounidis (eds.), *Decision Making and Knowledge Decision Support Systems*, Lecture Notes in Economics and Mathematical Systems 675, DOI 10.1007/978-3-319-03907-7_13, © Springer International Publishing Switzerland 2015

and value. The price is the result of the quantity agreed between the buyer and the seller. It is obvious that the price contains valuable information for calculating the value, but is not the best indicator of the intrinsic value [2]. The value is the estimation of the price that could have a company depending on multiple factors. This value depends on the expectations of the buyer or seller. At the same time, different buyers will have different expectations, so the valuation made will be also different for each other.

2 Preliminaries

2.1 Literature Review in Business Valuation

Although there is a vast academic literature on methods and models used in business valuation, the majority of the studies rest on a common scientific basis: the value of the company will be determined by the present value of cash flows that are expected in the future [3]. Nevertheless, some studies provide evidence that DCF are far from the market value of 10 % approximately. There are also studies demonstrating that models based on the benefit for action (BPA), adjust the proper value better than discounted cash flows (DCF).

In practice the DCF model has become the most popular, as it may seem that it is the most consistent with the objective of creating value, and encompasses the majority of factors that can influence the value of a company. It shows a more appropriate vision of the company value than methods based on the balance sheet [4].

There are studies [5] that show with numerical examples that the inclusion of the fuzzy philosophy in the DCF model gets a more accurate value and help investors to better measure their assets. Real options become the preferred methodology to minimize the effects of the uncertainty on academic literature, over traditional methods of valuation [6] or probabilistic approaches [7]. Subtle Sets are used to measure the value of goodwill, leaving behind the accounting methods, which do not correctly treat uncertainty, and thus determine a more adjusted the company's value [8].

Summarizing, the main limitations found in business valuation methods are that they are based on calculations between quantitative variables, but there are an infinite number of qualitative factors that influence the present value, which are not included in the valuation methods used. Fuzzy mathematics would allow us to include these qualitative variables in calculations to determine a tighter than its actual value of the company [9].

2.2   *Fuzzy Methodology in Business Valuation*

Research in the field of uncertainty has been an issue that for many decades has attracted increased interest since it is a fundamental and common in science concept. In 1965 was the first publication of Lofti A. Zadeh "Fuzzy sets" on fuzzy sets [10]. The initial idea that offered was a new logic allowing the binary logic (which can only take value 0 or 1), the multivalent logic (in which variables can take any score between 0 and 1 of the function characteristic of permanence). Against the Aristotelian principle of third except, prevailing in the processes of modeling for more than 2,000 years, the so-called "principle of simultaneity Gradual" [11], according to which a proposition can be both, true or false condition of assigning a value to the truth and falsehood value was imposed in 1996. In this way, it got a much more complete math that would make it possible to represent the information in a way most suited to the complex reality in which we live. From a business context and support the concept of [12] confidence interval, is possible because represent the inaccurate information in a more general way allowing to consider all the possible scenarios, from the position more optimistic to the most pessimistic in each of the variables considered. This has been useful, particularly to manage forecasts, especially ex-ante estimates, since usually the events are conditioned by subjective and uncertain elements.

3   Fuzzy Methodology for Uncertainty

To understand theories relating to uncertainty, we analyze basic concepts in the fuzzy methodology universe such as: confidence intervals, fuzzy numbers, triangular fuzzy numbers, fuzzy subset and probabilistic set, to finally reach to expertons [13].

3.1   *Confidence Intervals*

We will say that segment A is a confidence interval if we assume that the only information available in relation to a magnitude is greater or equal to a_1, and less than or equal to a_2 [14].

$$A = [a_1, a_2], \quad a_1 \leq a_2, \quad a_1, a_2 \in R. \tag{1}$$

3.2   *Fuzzy Numbers*

When we have a finite or infinite sequence of confidence intervals, which comply with the following properties:

1. Each confidence interval value will be $\alpha \in [0, 1]$, and won't be the same α value for any other. The resulting value is called presumption level.
2. The α level confidence interval $A_\alpha = [a_1^{(\alpha)}, a_2^{(\alpha)}]$ must comply:

$$\left(\alpha' < \alpha\right) \Rightarrow \left(A_\alpha \supset A_{\alpha'}\right), \quad \alpha, \alpha' \in [0, 1] \tag{2}$$

I.e. confidence intervals should fit strictly or not, with the other.
3. There is only one interval that can be reduced to a unique real.

In this case, the finite or infinite sequence of confidence intervals will be what we call by fuzzy number, which we represent by a capital letter, with the symbol——~ below, $\underset{\sim}{A}$. At the same time, the confidence level α interval will be appointed by A_α and we'll call it also "α-cut of $\underset{\sim}{A}$". We will say A_α is a functional application of α [15].

3.3   Triangular Fuzzy Numbers

The triangular fuzzy number (TFN), is represented by three numbers $a^{(1)}$, $a^{(2)}$ and $a^{(3)}$ as follows:

$$\underset{\sim}{A} = \left(a^{(1)}, a^{(2)}, a^{(3)}\right), \text{where } a^{(1)}, a^{(2)}, a^{(3)} \in R \text{ and } a^{(1)} \leq a^{(2)} \leq a^{(3)} \tag{3}$$

The essential feature of the TFN is that their μ functions are lineal:

$$\forall \alpha \in [0, 1] : A_\alpha = \left[\left(a^{(2)} - a^{(1)}\right)\alpha + a^{(1)}, -\left(a^{(3)} - a^{(2)}\right)\alpha + a^{(3)}\right] \tag{4}$$

3.4   Fuzzy Subsets

The theory of fuzzy subsets was defined by L.A. Zadeh (1965) [10]. Since then fuzzy subsets have become an ideal instrument to treat problems with the uncertainty data [16]. To define the concept of fuzzy subset, we consider a set or referential E and a regular subset A [17]. We say that for every point or element x of A it comply:

$$\mu_A(x) = 1 \text{ si } x \in A \qquad \mu_A(x) = 0 \text{ si } x \notin A \tag{5}$$

Where μ is the characteristic function of A, called "membership function".

The fuzzy subset is constructed when the elements of A, instead of taking only the value of 1 if $x \in A$, could also take a value a $\in [0, 1]$, including zero.

$$\mu_{\underline{A}}(x) = \alpha \in [0, 1] \tag{6}$$

A generalization of the concept of fuzzy set would be the probabilistic set.

3.5 Expertons

The concept of experton lies in two elements, the probabilistic set, and the confidence interval [18]. The merging of the two of them, originate the appearance of the experton, i.e. a probabilistic set, created through the merger of experts, but with opinions that have been expressed with intervals.

Any experton has no strict horizontal growing monotony (i.e. the bottom left is less than or equal to the upper right). Any experton has no strict vertical growing monotony (i.e. every number at level α' is larger or equal to every number at level α, if $\alpha < \alpha'$). Level 0 is always 1. Under mathematical formulation would be:

$$
\begin{aligned}
&1. \forall \alpha \in [0, 1] : a_1(\alpha) \leq a_2(\alpha) \text{ in } [a_1(\alpha), a_2(\alpha)] \\
&2. \forall \alpha, \alpha' \in [0, 1] : (\alpha' > \alpha) \rightarrow (a_1(\alpha) \leq a_1(\alpha'), a_2(\alpha) \leq a_2(\alpha')) \\
&3. (\alpha = 0) \rightarrow (a_1(\alpha) = 1, a_2(\alpha) = 1).
\end{aligned} \tag{7}
$$

4 Application on Business Valuation

When faced with business valuation, the financial information in the annual accounts is essential vital. It will form the basis of the creation of value for shareholders and will reflect the ability of the Organization to meet the goal which was created. However, there is other information that is not included in the financial statements even more important, such as the expectations. Financial information is very important, but do not complete all the necessary information when analyzing the business valuation. We want to show, in addition to the information which is not contained in the mandatory documents, within what we believe known financial information and voluntary information, there may be subjectivity. Uncertainty may affect the assessment of such information, since we are dealing with expected data which is not a certain reality.

Only the information in financial statements is certain [19]. The information derivate from them (auditory reports) and any other information from the company website, reports, presentations, etc., it's not included on the business valuation methods [20].

What we want to demonstrate is the lack of information on business valuation methods most widely used according to what we have found in the academic literature, the DCF method. We will discuss variables involved and that these variables may not be correct without taking into account the fuzzy logic introduced in the algorithms on this work.

5 Discounted Cash Flows

We will now focus on methods based on DCF. As we have found in the literature reviewed [21], the main methods used in valuing companies are based on discounting cash flows.

The basic formula for calculating the present value of these cash flows is based on the following expression:

$$\text{Company Value} = \sum_{i=1}^{n} \frac{CF_i}{(1+k)^i} + \frac{V_n}{(1+k)^n}. \tag{8}$$

CF are the expected cash flows; k discount rate applied; V_n is the residual value of the company at the end of the study period, and n is the number of years over which is expected to be generated income in the future. Therefore, those will be the factors that can affect business valuation according to the DCF method [22].

Some researchers [23] have found that under the DCF method the value could be overestimated. First for being too optimistic with the cash flows that will be generated in the future, this may cause an increase in the value. And the second would be to underestimate the cost of capital (discount rate). If we combine the two effects, the gap could be higher.

If we also add that neither is true the number of year over which is expected to be generating cash flows, the uncertainty of estimation is maximum.

Another point for the uncertainty is that this calculated period expected by the company, does not have to coincide with the actual period of value creation that will take the Organization in the market.

I would be remiss to these three variables add a fourth that would be the residual value of the company at the time n.

Below is a small study of in that the values of the four factors may vary if we don't apply any methodology that would reflect the subjectivity of each of the parameters.

5.1 n: Number of Years

As number of years in businesses valuation, we understand the time we consider to generate cash flows in the organization. The question is to quantify for how long the company will forecast future cash flows.

The time horizon of the projection, therefore, will depend on many factors, among which may be considered: stability expected in the economy, the industry and the company in particular, the variability of the macroeconomic variables, such as inflation, interest rate, exchange, taxation rates. The perception and experience of the financial analyst who develops the assessment also affect the given number of years. The type of organization and business is carried out, are also critical when evaluating the estimate of the exact number of periods will be taken into consideration [24].

5.2 CF: Cash Flow

To calculate cash flow, we use the information collected through the financial statements and adjusted to Free Cash Flows (FCF).

If we have to calculate these amounts for future periods, we will be working with a level of increasing uncertainty, considering that these are cash flows that are considered in the DCF formula.

5.3 k: Discount Rate

As a discount rate we should take the weighted average cost of capital (WACC). The formula is as follows:

$$WACC = \frac{D}{V}k_d(1-t) + \frac{E}{V}k_e. \tag{9}$$

The first part is the percentage of debt times the cost of debt considering the taxes. The second part of the formula is the percentage of equity times the cost of equity.

The percentage of debt and equity, are not constant throughout the life of the organization. If there is a change in the percentages, the value of the WACC will change, and affect the calculated value of the company.

The cost of debt (k_d) will not generate excessive problems, because it could be a known value and certain. The cost of equity (k_e) could be more difficult to calculate. The most common method for calculating the cost of capital is through the CAPM model (Capital Asset Pricing Model) [25].

5.4 V_n: Residual Value

In some cases the residual value of a company could mean more than 50 % of the total value, so it also constitutes as a fundamental variable in the analysis. Of the various theories about how to determine the residual value, the most used ones are [26]: book value of the company, liquidation value, future value of cash flow in perpetuity and future value of profits in perpetuity.

For its calculation we need to know the certain value of the previous variables, because we need the present value of this residual value.

6   Results and Conclusions

Once we have analyzed the four variables, now already we can express with more security the consequences on the reliability of the valuation if it will be done through the opinions expressed by the experts. The uncertainty detected in each of the variables would have been treated properly using fuzzy logic. This is why we believe that the inclusion of the fuzzy logic will enable future lines of research and development of traditional methods, providing them with a greater adaptability to business realities and the changing circumstances of the environment.

References

1. H. Barberis, M. Huang, T. Santos, Quart. J. Econ. **116**(1), 1–53 (2001)
2. S. Bhojraj, C. Lee, J. Account. Res. **40**(2), 407–439 (2002)
3. L. Gonzalez Jimenez, L. Blanco Pascual, Eur. J. Finance **16**(1), 57–78 (2010)
4. T. Copeland, T. Koller, J. Murrin, *Measuring and Managing the Value of Companies*, 2nd edn. (Makinsey, New York, NY, 2000)
5. J. Yao, M. Chen, H. Lin, Expert Syst. Appl. **28**(2), 209–222 (2005)
6. I. Ucal, C. Kahraman, Ūkio Technologinis Ir Ekonominis Vystymas **15**(4), 646–669 (2009)
7. C. Carlsson, R. Fuller, Fuzzy Set. Syst. **139**(2), 297–312 (2003)
8. I. Ionita, M. Stoica, J. Econ. Forecast. **10**(1), 115–122 (2009)
9. A.M. Gil-Lafuente, in *Studies in Fuzziness and Soft Computing*, vol. 287 (2012), pp. 177–189
10. L. Zadeh, Inform. Control **8**(3), 338 (1965)
11. J. Gil Aluja, in *Proceedings Del III Congreso SIGEF*, Buenos Aires, Noviembre (1996)
12. A.M. Gil-Lafuente, *Fuzzy logic in financial analysis* (Springer, Berlin, 2005)
13. A. Kaufmann, J. Gil Aluja, *Técnicas operativas de gestión para el tratamiento de la incertidumbre* (Editorial Hispano Europea, Spain, 1987)
14. A. Kaufmann, J. Gil Aluja, *Colección: Economía y Administración de Empresas* (Editorial Piramide, Spain, 1992)
15. A. Kaufmann, J. Gil Aluja, *Técnicas especiales para la gestión de expertos* (Editorial Milladoiro, Spain, 1993)
16. A. Kaufmann, Fuzzy Set. Syst. **28**(3), 295–304 (1988)
17. A. Kaufmann, *Introduction to the Theory of Fuzzy Subsets* (Academic, New York, NY, 1975)
18. S. Imam, R. Barker, C. Clubb, Eur. Account. Rev. **17**(3), 503–535 (2008)
19. D. Audretsch, A. Link, Small Bus. Econ. **38**(2), 139–145 (2012)
20. A.M. Gil-Lafuente, L. Barcellos, CSEDU Proceedings of the Second International Conference on Computer Supported Education, Valencia, Spain (2010)
21. B. Hrvol'ova, J. Markova, L. Nincak, Ekonomický c'Asopis **59**(2), 148–162 (2011)
22. P. Fernández, *Gestión 2000* (Terceraedición, Spain, 2005)
23. R. Knight, M. Bertoneche, *Financial Performance* (Reed Educational and Professional Publishing Ltd England, 2001)
24. T. Garicano, Harvard-Deusto Finanzas y Contabilidad **59**, 12–21 (2004)
25. J. Tobin, Rev. Econ. Stud. **25**(66–6), 65–86 (1957)
26. I. Martínez Conesa, E. García Meca, *Valoración de empresas cotizadas* (AECA, España, 2005)

An Efficient ANFIS Based Approach for Screening of Chronic Obstructive Pulmonary Disease from Chest CT Scans with Adaptive Median Filtering

K. Meenakshi Sundaram and C.S. Ravichandran

Abstract Medical diagnostic and imaging system are ubiquitous in modern health care facilities. The advantages of early detection of potential lesions and suspicious masses within the bodily tissue have been well established. It can be detected and assessed many different types of injuries, diseases, and conditions with the aid of the medical imaging that allows medical personnel to look into living cells non-instructively. Chronic Obstructive Pulmonary Disease (COPD) is the fourth leading cause of death worldwide and the only chronic disease with increasing mortality rates. COPD is the name for a group of lung diseases including chronic bronchitis, emphysema and chronic obstructive airways disease. This paper involves in improving the accuracy over the existing technique using the adaptive region growing property and Adaptive-Neuro-Fuzzy Inference System-ANFIS classifier. Initially, pre-processing is carried out for the input image by Adaptive median Filter technique to make the image suitable for further processing. The contours of the image will be obtained using region growing technique. The ANFIS classifier is then used to confirm the suspected COPD cavities. The classification will be carried out by the features which have been taken from the segmented image. The proposed technique is implemented in MATLAB and the performance is compared with the existing technique. From the experimental result it can be said that the proposed method achieved more accuracy as compared with existing techniques.

Keywords Adaptive-Neuro-Fuzzy Inference System-ANFIS • Emphysema and chronic obstructive airways disease

K. Meenakshi Sundaram (✉)
Department of Information Technology, Dhanalakshmi Srinivasan College of Engineering, Coimbatore, Tamilnadu, India
e-mail: meenaksji@gmail.com

C.S. Ravichandran
Department of Electrical & Electronics Engineering, Sri Ramakrishna Engineering College, Coimbatore, Tamilnadu, India

A.M. Gil-Lafuente and C. Zopounidis (eds.), *Decision Making and Knowledge Decision Support Systems*, Lecture Notes in Economics and Mathematical Systems 675, DOI 10.1007/978-3-319-03907-7_14, © Springer International Publishing Switzerland 2015

1 Introduction

The Medical imaging is one of the most useful diagnostic tools available in modem medicine. Medical diagnostic and imaging system are ubiquitous in modern health care facilities. The advantages of early detection of potential lesions and suspicious masses within the bodily tissue have been well established. COPD is a most important disease but under-recognized cause of morbidity and mortality worldwide [1]. The occurrence of COPD in the general population is predictable to be, ~1 % across all ages rising steeply to >10 % amongst those aged 40 years. The occurrence climbs appreciably higher with age. The 30-year for the global increase in COPD is from 1990 to 2020 is surprising. COPD is projected to move from the sixth to the third most common cause of death worldwide, at the same time as rising from fourth to third in terms of morbidity within the same time-frame [2]. The cofactors in charge for this extraordinary increase are the continued use of tobacco, coupled with the changing demographics of the world, such that many more people, especially those in developing countries, are living into the COPD range of age.

COPD occurrence is generally higher than recognized by health authorities [3, 4]. Few population-based occurrence surveys have been carried out, and prevalence estimates have repeatedly relied on expert opinion or self-reported doctor diagnosis, a disreputably unpredictable source of information for COPD. For example, in the USA National Health and Nutrition Examination Survey III, 70 % of those with airflow obstruction had never received the diagnosis of COPD [5]. The IBERPOC study in Spain also reported that there was no previous diagnosis of COPD in 78 % of identified cases and, even more worrisome, only 49 % of those with severe COPD were receiving some kind of treatment for COPD [6]. Recently, the Nippon COPD Epidemiology (NICE) study in Japan, presented the current series, had a similar finding [7].

During the 1990s, surveys of asthma successfully identified a huge variations in asthma occurrence in children and adults, as high as 20-fold. It shows that the geographical distribution of COPD is more harmonized than asthma, at least in the developed countries. It seems that the distribution of COPD follows the distribution of its risk factors very closely, of which smoking is undoubtedly the most important worldwide. COPD is in the spotlight of worldwide at the high prevalence, morbidity and mortality present challenges for healthcare systems. From the patient's viewpoint, it is also a disease that has a reflective effect on quality of life [8]. The burden of COPD can be assessed in a number of ways such as mortality, morbidity, prevalence, disability-adjusted life years, cost and quality of life. A number of authors have reviewed this topic in detail elsewhere [9–11].

In this paper, first the input image is pre-processed; the lung region is segmented from that image, segmented the cavity region in that lung region, extracted some features for training the classifier and used the adaptive neuro-fuzzy inference system (ANFIS) classifier to identify the tuberculosis affected lung. The pre-processing is done by using the Adaptive Median Filter to avoid the noise in the input image and to increase the image quality. The cavity segmentation is done by evaluating the pixel range in the segmented lung region and setting a threshold value from that evaluated pixels and comparing every pixel with that threshold value. After the lung and cavity segmentation, some parameters are chosen to train the classifier to identify whether an X-ray image is a normal or tuberculosis affected. The classifier used in proposed technique is ANFIS classifier. The ANFIS Classifier is then trained using the parameters chosen from the sample chest CT scan images to identify the normal lung and tuberculosis affected lung.

2   Related Work

The computer-aided diagnosis (CAD) system is used for early detection of tuberculosis in lungs by analyzing chest 3D computed tomography (CT) images. The underlying idea of developing a CAD system is not to delegate the diagnosis to a mechanism, but quite than a machine algorithm acts as a support to the radiologist and points out locations of suspicious objects, so that the overall sensitivity is raised. CAD systems meet four main objectives, which are improving the quality and accuracy of diagnosis, increasing therapy success by early detection of cancer, avoiding unnecessary biopsies and reducing radiologist interpretation time [12].

V.M. Katoch [13] explained that the diagnosis of tuberculosis is mostly based on clinical features, demonstration of acid fast bacilli (AFB) and isolation of Mycobacterium tuberculosis from the clinical specimens. These techniques have limits of speed, warmth and specificity. Several rapid techniques for detection of early growth have been described for last two decades which can help in obtaining the culture and sensitivity reports early. Important among such methods are BACTEC, mycobacterial growth indicator tuber (MGIT), and Septi-chek, MB/BacT systems. This growth can be recognized by rapid methods based on lipid analysis and specific gene probes, PCR-RFLP methods and ribosomal RNA sequencing. Advance improvement in knowledge about genetic structure of tubercle bacillus helps to develop gene probes and gene amplification methods for detection of tubercle bacillus, from culture or directly in clinical specimens and molecular detection of drug resistance. The gene probes can help in rapid identification of isolates, gene amplification methods developed for diagnosis of tuberculosis are obviously highly sensitive especially in culture negative specimens from different paucibacillary forms of disease. The molecular methods drug resistant mutants for drugs like rifampicin can be detected with reasonable certainty within hours.

In recent years great advances have been made in Computer Aided Diagnosis (CAD) systems for detecting disease from Computed Tomography scans, mainly due to the advances made in the scanning machines which allow a greater amount and quality of information to be extracted during a single breath of the patient. The use of textural analysis and pattern recognition techniques for regression or classification is most suited to the evaluation of global conditions (e.g. Ground glass, Emphysema) rather than local (small nodules), which won't be concerned with in this report. This recent progress in CAD in Chest Radiology has been discussed in

Giger [14], where it has been noted that the amount of 3-D image data from thoracic CT scans greatly increases the number of images that much be reviewed by the radiologist and therefore a search aid may be a great benefit.

Uchiyama et al. [15] said that the selected regions in 315 HRCT images from 105 patients, relating to six different patterns, i.e., ground-glass opacities, reticular and linear opacities, nodular opacities, honeycombing, emphysematous change, and consolidation, labeled by three radiologists. The lungs were first segmented, using standard technique, then divided into many contiguous regions of interest (ROIs) with a 32×32 matrix and classified using artificial neural networks. The accuracy varied from 88 to 100 %, with specificity in detecting a normal ROI of 88.1 %.

Sluimer et al. [10, 16] presented a CAD system to automatically distinguish normal from abnormal tissue in HRCT chest scans of 116 patients, producing 657 ROIs labeled as containing normal or abnormal tissue. The circular ROIs with an 80-pixel diameter were extracted from the peripheral lung region in slices at the height of the aortic arch, with each ROI required to contain at least 75 % abnormal tissue. An accuracy of 86.2 % as obtained, comparable to those of a radiologist when evaluating only the ROIs, i.e. without seeing the whole scan.

Lee et al. [17] explained that the digital CT is a technique used for recording images in computer code as an alternative of on X-ray film. The images are displayed on a computer monitor can be enhanced or lightened or darkened before they are printed on film. Images can also be manipulated; the radiologist can magnify or zoom in on an area. This screening will generate large number of CT images to be determined by a small number of radiologists resulting in misdiagnosis due to human errors caused by visual fatigue. The sensitivity of human eye decreases with increasing number of images. Hence, it may be helpful for a radiologist, if a computer-aided system is used for detection of tumours in CT images. Computer-aided detection (CAD) involves the use of computers to bring suspicious areas on a CT to the radiologist's attention. It is used after the radiologist has done the initial review of the CT. There are several image processing methods proposed for extract of tumours from CT image for better view of area and shape of tumour.

3 Proposed Technique for the Identification of Cavity

The block diagram of the proposed approach is shown in Fig. 1. In this figure some sample chest CT scan images are taken with COPD and without COPD. The sample images are then preprocessed and then send for segmentation process. There segmenting the lung and cavity regions. After the lung and cavity regions are segmented from the sample images, some parameters are chosen to train the classifier. First the preprocessing is done to find weather the COPD is affected or not. After the preprocessing process, need to segment the lung and the cavity region. After that the chosen parameters are given to the classifier, here the

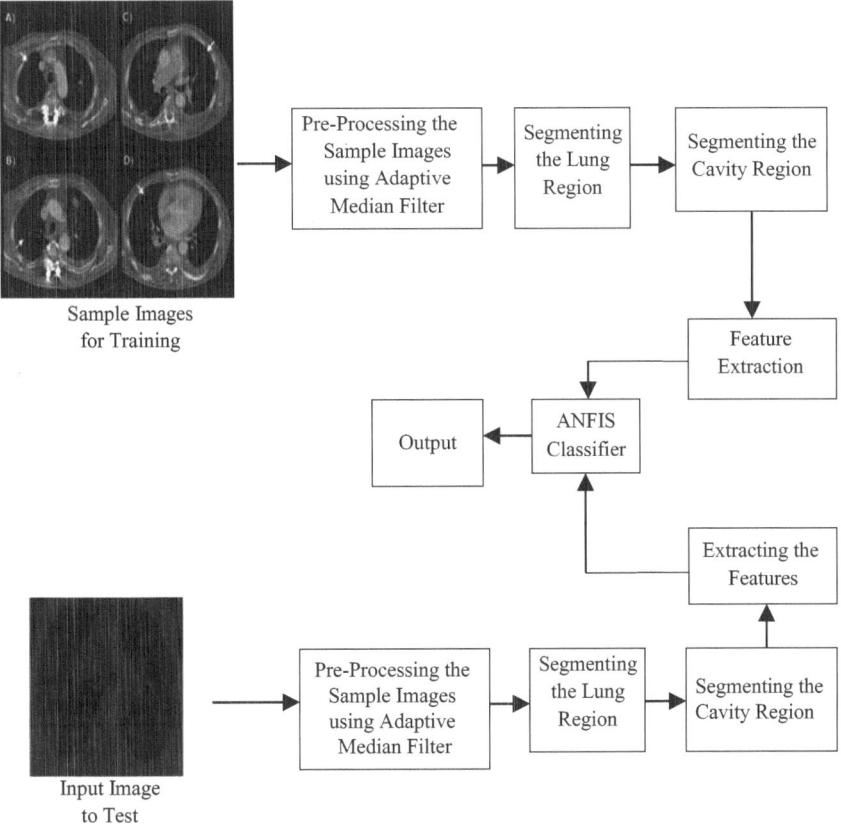

Fig. 1 Block diagram for proposed technique

ANFIS classifier is used. The ANFIS classifier then identify whether the input chest CT scan image is affected by COPD or not by comparing the parameters from the sample images and from the input image.

3.1 Pre-processing

The input image is subjected to the pre-processing steps to make the image suitable for further process. The pre-process is used to load the input image to the MATLAB environment and it will remove the noise present in the input image. The image is passed through the Adaptive median filter to lower the noise and to get a better image. The adaptive median filter will also increase the image quality and the corner of the images.

Adaptive Median Filter The standard median filter performs well as long as the spatial noise density of the salt and pepper noise is not large. The filter performance degrades when the spatial noise variance of the salt and pepper noise increases Chen and Whu (1998). Further with larger image and as the size of the kernel increases, the details and the edges become obscured Maragos and Schafer (1987). The standard median filter does not take into account the variation of image characteristics from one point to another. The behavior of adaptive filter changes based on statistical characteristic of the image inside the filter region defined by the mxn rectangular window S_{xy} Maragos and Schafer (1987). The adaptive median filter differs from other adaptive filter as the size of the rectangular window S_{xy} is made to vary depending on:

- $Z_{min} = $ Minimum gray level value in S_{xy}
- $Z_{max} = $ Maximum gray level value in S_{xy}
- $Z_{med} = $ Median of gray level in S_{xy}
- $Z_{xy} = $ Gray levels at coordinate (x, y)
- $S_{max} = $ Maximum allowed size of S_{xy} Gonzalez and Woods (2002).

The flowchart of adaptive median filtering is based on two levels is shown in the Fig. 2.

The adaptive median filtering algorithm works in two levels, denoted by LEVEL1 and LEVEL2. The application of AMF provides three major purposes: to denoise images corrupted by salt and pepper (impulse) noise; to provide smoothing of non-impulsive noise, and also to reduce distortion caused by excessive thinning or thickening of object boundaries. The values Zmin and Zmax are considered statistically by the algorithm to be 'impulse like' noise components, even if these are not the lowest and highest possible pixel values in the image.

The purpose of LEVEL1 is to determine if the median filter output Zmed is impulse output or not. If LEVEL1 does find an impulse output then that would cause it to branch to LEVEL2. Here, the algorithm then increases the size of the window and repeats LEVEL1 and continues until it finds a median value that is not an impulse or the maximum window size is reached, the algorithm returns the value of Zxy. Every time the algorithm outputs a value, the window Sxy is moved to the next location in the image. The algorithm is then reinitialized and applied to the pixels in the new location. The median value can be updated iteratively using only the new pixels, thus reducing computational overhead.

Two commonly used small kernels are shown in Fig. 3.

Because these kernels are approximating a second derivative measurement on the image, they are very sensitive to noise. To counter this, the image is often smoothed before applying the Adaptive median filter. This pre-processing step reduces the high frequency noise components prior to the differentiation step.

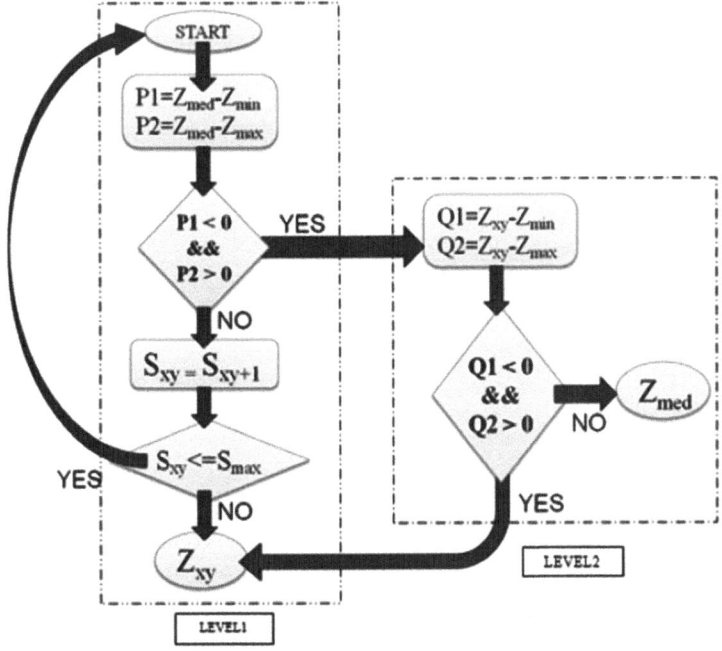

Fig. 2 Flowchart of adaptive median filter

0	-1	0
-1	4	-1
0	-1	0

-1	-1	-1
-1	8	-1
-1	-1	-1

Fig. 3 Two commonly used discrete approximations to the adaptive median filter

3.2   Lung Segmentation

Lung segmentation is a process of segmenting the lungs from the chest CT scan image. The normal process of region growing technique for segmenting the lungs is shown in the Fig. 4. First choose a pixel from the chest CT scan image as default. Then need to set a threshold value for comparison to find the pixel intensity for the lung area in the chest CT scan. The default pixel which chosen is compared with the adjacent pixel values. If the difference between the default pixel and the adjacent pixel is greater than the threshold value, have to exclude that adjacent pixel. If the

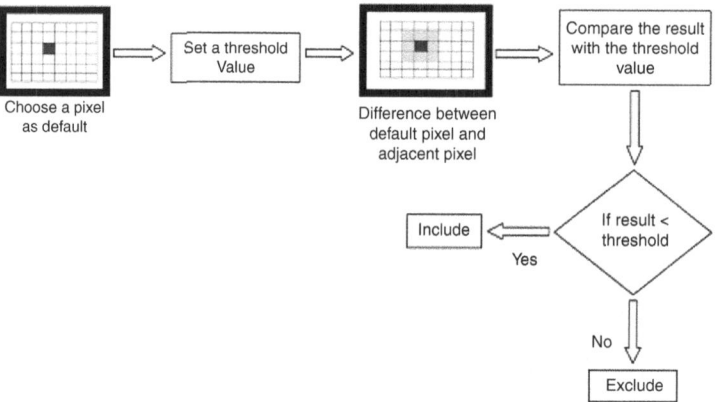

Fig. 4 Block diagram of normal region growing technique

difference between the default pixel and the adjacent pixel is less than the threshold value, have to include that adjacent pixel for region growing. Compare all the pixels except the left pixels with its adjacent pixels by keeping one pixel as default. The process of normal region growing technique is shown in the Fig. 4.

In this paper, comparing the normal region growing technique with the Local Gabor XOR Pattern (LGXP) based region growing technique to segment the lungs from the chest CT scan image. The LGXP technique is used to find the texture image.

The LGXP based region growing technique is as follows. In LGXP technique, apply the Gabor Phase Technique on every pixel in the chest CT scan image. The Gabor Phase Technique will convert all the pixel values to phase values (0–360). After converting all the pixel values to phase values, find these phase values comes under which quadrant. Each quadrant has certain values. For the first quadrant the value is zero and for the second quadrant the value is one and for the third quadrant the value is two and for the fourth quadrant the value is three. After that choose a default phase value of a pixel and check under which quadrant this phase value comes and assign respective quadrant value to that pixel. After assigning respective quadrant value to the default pixel, check the adjacent pixel's phase values and assign the respective quadrant values to those adjacent pixels. Then convert the adjacent pixel's value as zero which has the same quadrant value of the default pixel. If the adjacent pixels value does not have the same quadrant value of the default pixel, convert the adjacent pixel's value as one. Now the pixel values would be like binary values as zeros and ones. After converting the pixel values as binary format, make that binary format as decimal value and apply that decimal value to the default pixel. The process of taking the binary value is shown in the figure. Likewise apply this LGXP process for all the pixels in the chest CT scan by keeping one pixel as default. The sample process of LGXP technique is shown in the Fig. 5.

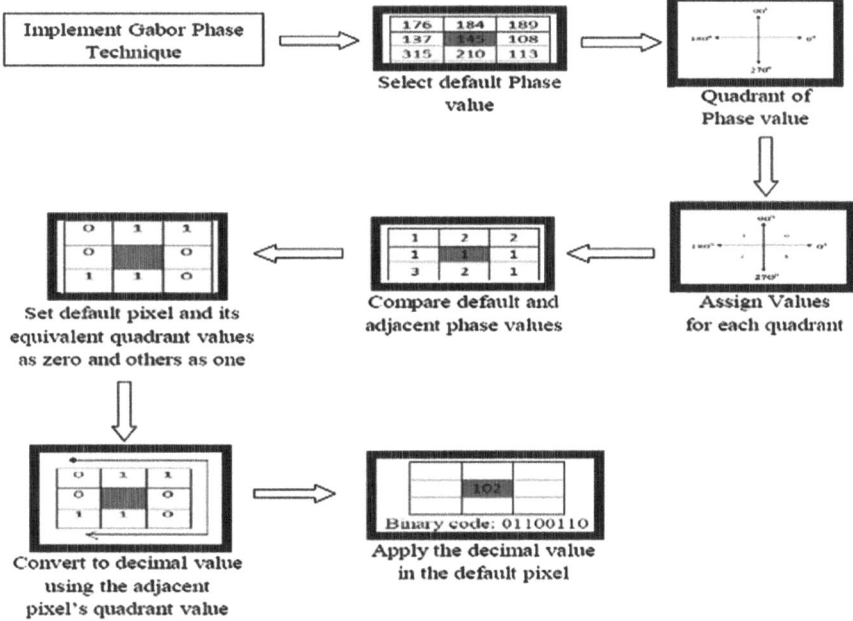

Fig. 5 Block diagram of LGXP Technique

3.3 Local Gabor XOR Pattern

The fundamental idea of technique [18] is to ease the sensitivity of Gabor phase to the differing positions, whether two phases reflect same local feature must be determined in a "looser" way. Specifically, if two phases belongs to the same interval (for instance: 00, 900), they are believed to have similar local features or else they reflect different local features. In this section, the LGXP descriptor is presented.

The Fig. 6 shows an instance for the LGXP encoding method where the phase is quantized into four ranges. In LGXP technique, phases are first quantized into disparate ranges and the LGXP operator is applied to the quantized pixels of the central pixel and each of its neighboring pixels and eventually the result of the binary labels are concatenated together as a local pattern of the central pixel. In the Fig. 4, (a) is the matrix with initial phase of the pixels after applying the Gabor filter and (b) is the result after quantization and (c) is the result after XOR comparison with the center quantized value. From the matrix which we got after XOR comparison, we can deduce the binary value obtained is 01011101 and its equivalent decimal value is 93. The pattern of LGXP in binary and decimal form is as follows:

Fig. 6 Example of LGXP method where the phase is quantized into four ranges

95	32	14
21	13	78
15	18	25
(a)		

1	3	1
2	1	0
1	2	2
(b)		

0	1	0
1	0	1
0	1	1
(c)		

$$LGXP_{\mu v}(P_c) = \left[LGXP_{\mu v}^N, LGXP_{\mu v}^{N-1}, \ldots, LGXP_{\mu v}^1 \right]_{binary} \left[\sum_{i=1}^{N} 2^{i-1}.LGXP_{\mu v}^1 \right]_{decimal} .$$

$$(1)$$

where, P_c denotes the central pixel in the Gabor phase map with scale v and orientation μ, N is the size of the neighborhood and $LGXP_{\mu v}^i (i = 1, 2, \ldots, N)$ denotes the pattern calculated between P_c and its neighbor P_i, which is computed as follows:

$$LGXP_{\mu v}^i = q(\Phi_{\mu v}(P_c)) \otimes q(\Phi_{\mu v}(P_i)), \quad i = 1, 2, \ldots N. \tag{2}$$

where $\Phi_{\mu v}$ denotes the phase, $\otimes$ denotes the LXP operator, which is based on XOR operator, q denotes the quantization operator which calculates the quantized code of the phase according to the number of phase ranges.

$$a \otimes b = \begin{cases} 0, & \text{if } a = b \\ 1, & \text{else} \end{cases}$$

$$q(\varnothing_{\mu v}(.)) = i, \quad \text{if } \frac{360 * i}{e} \leq \Phi_{\mu v}(.) < \frac{360 * (i+1)}{e}, \tag{3}$$

$$i = 0, 1, \ldots \ldots b - 1.$$

where, e denotes the number of phase ranges. With the pattern explained above, one pattern map is computed for each Gabor kernel. Thereafter, each pattern map is split into m non overlapping sub blocks and the histograms of all the sub blocks of scales and the orientations are concatenated to form the proposed LGXP descriptor of the input face image

$$H = \left[H_{\mu_0 v_0 1}, \ldots \ldots, H_{\mu_0 v_0 m_i}, \ldots \ldots, H_{\mu_{O-1} v_{s-1} 1}, \ldots \ldots, H_{\mu_{O-1} v_{s-1} m} \right]. \tag{4}$$

where $H_{\mu v}$ ($i = 1, 2, \ldots, m$) denotes the histogram of the ith sub block of the LGXP map with scale v and orientation.

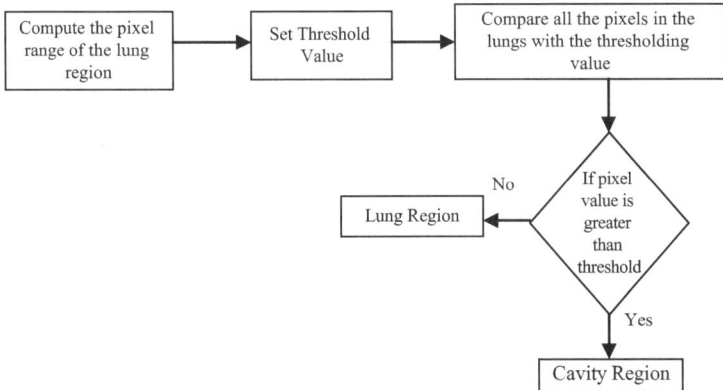

Fig. 7 Block diagram of cavity segmentation

3.4   Cavity Segmentation

After the lung segmentation, identify the cavities in the lung. The cavities present in the lung region are an essential thing to identify the COPD affected lung. To identify the cavity in the lung, set an adaptive threshold value. The threshold value is chosen by calculating the pixel range in the lung region and dividing that pixel range by two. After that, compare the threshold with all the pixels. While comparing the pixels to the threshold value, if the pixel value is greater than the threshold value then it would be the cavity region and if the pixel value is less than the threshold value then it would be the lung region. The Fig. 7 shows the block diagram for segmenting the cavity region from the lung region.

3.5   Feature Extraction

After finding the regions, extract some features to diagnose the disease in the lung. To discover the disease in the lung, have to feed the extracted feature into the classifier, because the extracted features will give vital information about the region which is used to train the classifier. In this paper an ANFIS classifier is used for feature extraction. The features need to extract are number of cavities in the lung region, minimum area of cavity region, maximum area of cavity region, total number of pixels in each cavity, maximum repeated pixel intensity in the cavity region and maximum repeated pixel in the lung region to find the total number of cavities in the lung region. Because the normal lung would also have some cavities present in its region. So to distinguish the normal lung image and the COPD affected lung should find the total numbers of cavities present in the lung region and give the result to the ANFIS classifier.

3.5.1 Adaptive Neuro-fuzzy Inference System Classifier

Architecture of ANFIS

The ANFIS is a fuzzy Sugeno model put in the structure of adaptive systems to make easy learning and adaptation Jang (1993). Such structure makes the ANFIS modeling more efficient and less reliant on expert knowledge. To present the ANFIS structural design, two fuzzy if–then rules based on a first order Sugeno model are measured:

Rule 1: If (x is A_1) and (y is B_1) then ($f1 = p_1x + q_1y + r_1$).
Rule 2: If (x is A_2) and (y is B_2) then ($f2 = p_2x + q_2y + r_2$),

where x and y are the inputs, A_i and B_i are the fuzzy sets, f_i are the outputs within the fuzzy region precise by the fuzzy rule, pi; qi and ri are the design parameters that are determined throughout the training process. The ANFIS architecture to put into practice these two rules is shown in Fig. 8, in which a circle indicates a fixed node, while a square indicates an adaptive node.

In the first layer, all the nodes are adaptive nodes. The outputs of layer 1 are the fuzzy membership grade of the inputs, which are given by:

$$O_i^1 = \mu_{Ai}(x), \quad i = 1, 2, \tag{5}$$

$$O_i^1 = \mu_{Bi-2}(y), \quad i = 3, 4. \tag{6}$$

where $\mu_{Ai}(x)$, $\mu_{Bi-2}(y)$ can adopt any fuzzy membership function. For example, if the bell shaped membership function is employed, $\mu_{Ai}(x)$ is given by:

$$\mu_{Ai}(x) = \frac{1}{1 + \left\{ \left(\frac{x-c_i}{a_i} \right)^2 \right\} b_i.} \tag{7}$$

where a_i, b_i and c_i are the parameters of the membership function, governing the bell-shaped functions, as a result.

In the second layer, the nodes are fixed nodes. They are labeled with M, indicating that they carry out as a simple multiplier. The outputs of this layer can be correspond to as:

$$O_i^2 = w_i = \mu_{Ai}(x)\mu_{Bi}(y), \quad i = 1, 2. \tag{8}$$

which are the called as firing strengths of the rules.

In the third layer, the nodes are also fixed nodes. They are labeled with N, indicating that they play a normalization role to the firing strengths from the preceding layer.

The outputs of this layer can be represented as:

Fig. 8 ANFIS architecture

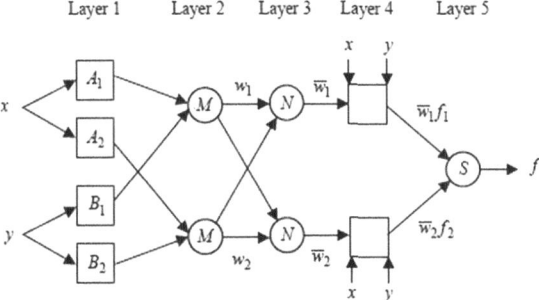

$$O_i^3 = \overline{w}_i = \frac{w_i}{w_1 + w_2}, \quad i = 1, 2. \tag{9}$$

which are the so-called normalized ring strengths.

In the fourth layer, the nodes are adaptive nodes. The output of each node in this layer is simply the product of the normalized firing strength and a first-order polynomial. Thus, the outputs of this layer are given by:

$$O_i^4 = \overline{w}_i f_i = \overline{w}_i(p_i x + q_i y + r_i), \quad i = 1, 2. \tag{10}$$

In the fifth layer, there is only one single fixed node labeled with S. This node performs the summation of all incoming signals. Hence, the overall output of the model is given by:

$$O_i^5 = \sum_{i=1}^{2} \overline{w}_i f_i = \frac{\left(\sum_{i=1}^{2} w_i f_i\right)}{w_1 + w_2}, \quad i = 1, 2. \tag{11}$$

It can be experiential that there are two adaptive layers in this ANFIS structural design, that is the first layer and the fourth layer. In the first layer, there are three changeable parameters $\{a_i, b_i, c_i\}$ which are connected to the input membership functions. These parameters are the so-called premise parameters. In the fourth layer, there are also three modifiable parameters $\{p_i, q_i, r_i\}$, pertaining to the first order polynomial. These parameters are so-called consequent parameters [12, 13].

4 Training and Testing Using ANFIS Classifier

Some of the data features are to be taken to identify the normal lung region and COPD affected lung by this the classifier is trained. The data features will then train the classifier and the classifier will find whether the given CT scan image is normal or abnormal. The data features which have chosen for training the ANFIS classifier are number of cavities in the lung region, maximum area of cavity in the lung region, minimum area of cavity in the lung region, total number of pixels in each cavity, maximum repeated pixel in the cavity regions together and maximum repeated pixel in the lung region. After computing all the data features, have to give the values to the classifier. For instance, choosing three normal CT scan images and three abnormal CT scan images, need to calculate all the six data features separately for all the CT scan images had chosen. After calculating all the six data features for every chosen CT scan images, have to give the result to the ANFIS classifier. Using those results train the classifier to identify the normal and abnormal lung from the given CT scan image. After the ANFIS classifier is trained, give a new CT scan image to find whether it has COPD or not. Afterwards, the six data features such as number of cavities in the lung region, maximum area of the cavity region, minimum area of the cavity region, total number of pixels in each cavity, maximum frequent pixel in the cavity region and maximum repeated pixel in the lung region are calculated for the new CT scan image. The computed values of all the six data features are then give to the ANFIS classifier. The ANFIS classifier is then comparing the values of all the six data features with the stored values of normal and abnormal CT scan images. Because during training have stored all the six data features of the five normal CT scan images and five abnormal CT scan images. After comparison, the ANFIS classifier will identify whether the given CT scan image comes under normal category or abnormal category.

5 Results and Discussion

The experimental result is conducted in MATLAB. The Fig. 9 shows the normal and abnormal lung images.

The sample images are taken and the images are filtered using Adaptive Median filter. The filtering technique is used to remove the noises and it improves the quality of the images as shown in the Fig. 10.

The sample images after applied the filtering technique are given to the process of lung segmentation. The lung segmentation process only segments the lung region from the sample CT scan images. The Fig. 11 shows a sample image of segmented lungs with COPD and without COPD.

After the lung is segmented from the sample images, have to segment the cavities from the lung region. Using the cavities in the lung region identify whether a lung is COPD affected or not. The Fig. 12 shows a sample image of segmented cavities and segmented cavities with CT scan image for the COPD affected lung.

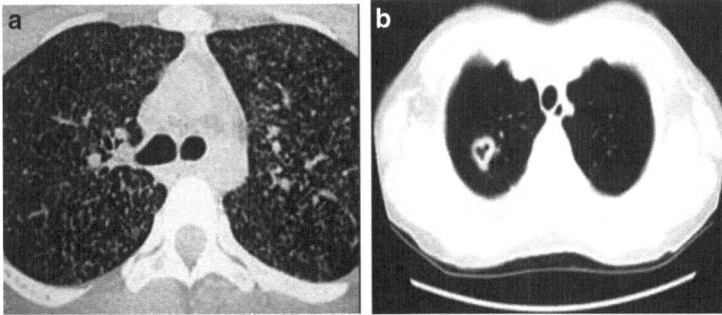

Fig. 9 Sample images of normal and abnormal lungs images. (**a**) Normal image. (**b**) Abnormal image

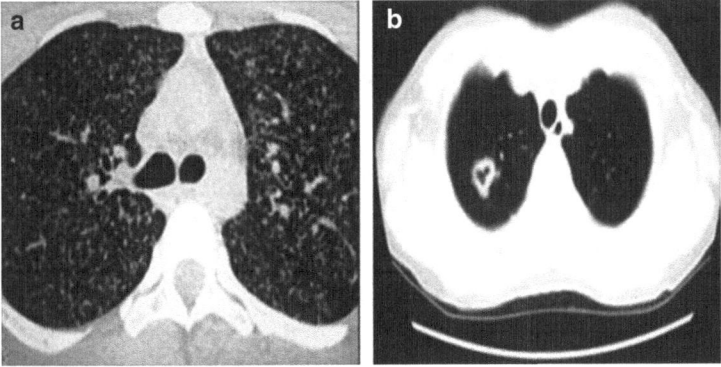

Fig. 10 Sample image of lungs after filtering process. (**a**) Normal image after filtering process. (**b**) Abnormal image after filtering process

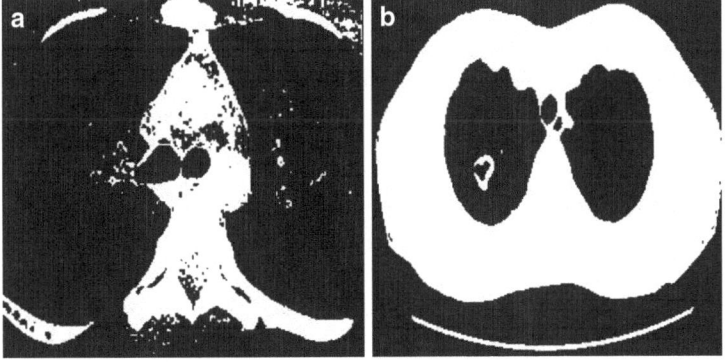

Fig. 11 Sample image of lungs after segmentation process. (**a**) Normal image after segmentation process. (**b**) Abnormal image after segmentation process

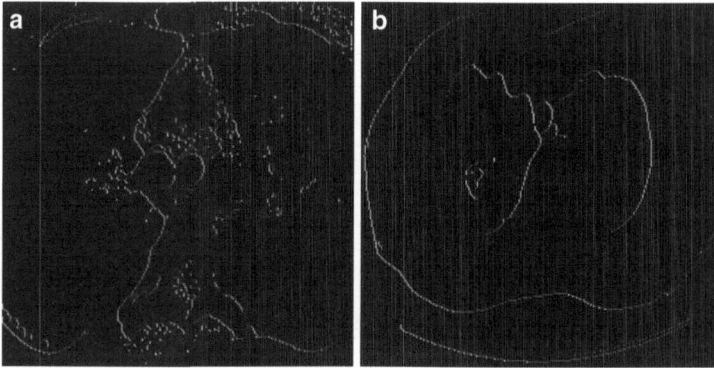

Fig. 12 Sample image of lungs after segmenting the cavities. (**a**) Normal image after segmenting cavities. (**b**) Abnormal image after segmenting cavities

6 Performance Analysis Using Evaluation Metrics

The evaluation of the COPD identification of the images is carried out using the following metrics:

$$Sensitivity = TP/(TP + FN) \tag{12}$$

$$Accuracy = (TN + TP)/(TN + TP + FN + FP) \tag{13}$$

where True Positive TP, True Negative TN, False Negative FN, False Positive FP.

Sensitivity is the proportion of true positives that are correctly identified by a diagnostic test. It shows how good the test is at detecting a disease.

Accuracy is the proportion of true results, either true positive or true negative, in a population. It measures the degree of veracity of a diagnostic test on a condition.

Table 1 shows the accuracy comparison between proposed technique and the existing technique. The tabular column shows that the proposed technique gives better performance than the existing technique.

7 Conclusion

In this paper, proposed an efficient technique for the detection of COPD in the lungs using CT scan Images. The proposed technique contains pre-processing, lung segmentation, cavity segmentation, feature extraction, training and testing using ANFIS classifier. The ANFIS classifier is efficient and simple in nature. The performance of the proposed technique and the existing technique is analyzed using evaluation metrics. To evaluate these metrics, should need some terms like True Positive, True Negative, False Positive and False Negative. After evaluating

Table 1 Comparative analysis of existing technique with proposed technique

Techniques	TP	TN	FP	FN	Sensitivity	Accuracy
SVM Technique	3	9	1	1	0.9	0.857
Proposed ANFIS	1	4	1	0	0.43	0.989

these metrics it shows that the performance of proposed technique is better when compared to the existing technique in terms of accuracy. The result shows that the accuracy of proposed technique higher than existing techniques.

References

1. R.A. Pauwels, K.F. Rabe, Lancet **364**, 613–620 (2004)
2. C.J. Murray, A.D. Lopez, Lancet **349**, 1269–1276 (1997)
3. R.A. Pauwels, A.S. Buist, P.M. Calverley, C.R. Jenkins, S.S. Hurd, Am. J. Respir. Crit. Care Med. **163**, 1256–1276 (2001)
4. GOLD Guidelines 2003. http://www.goldcopd.com (2005)
5. P. Stang, E. Lydick, C. Silberman, A. Kempel, E.T. Keating, Chest **117**, 354S–359S (2000)
6. V. Sobradillo-Pena, M. Miravitlles, R. Gabriel et al., Chest **118**, 981–989 (2000)
7. Y. Fukuchi, M. Nishimura, M. Ichinose et al., Respirology **9**, 458–465 (2004)
8. S. Rennard, M. Decramer, P.M. Calverley et al., Eur. Respir. J. **20**, 799–805 (2002)
9. G. Viegi, A. Scognamiglio, S. Baldacci, F. Pistelli, L. Carrozzi, Respiration **68**, 4–19 (2001)
10. S. Loveymi, B. Shadgar, A. Osareh, Image Fusion **6** (2011), pp. 5
11. M.H. Malik, S.A.M. Gilani, Anwaar-ul-Haq, J. Image Vis. **62** (2008)
12. B. Magesh, P. Vijayalakshmi, M. Abirami. Int. J. Comput. Trends Technol. May-June (2011)
13. V.M. Katoch, Central JALMA Institute for Leprosy & Other Mycobacterial Diseases (ICMR) (2003)
14. M.I. Giger, N. Karssemeijer, S.G. Armato, Guest editorial computer-aided diagnosis in medical imaging (2001)
15. Y. Uchiyama, S. Katsuragawa, H. Abe, J. Shiraishi, F. Li, Q. Li, C.-T. Zhang, K. Suzuki, K. Doi, Med. Phys. **2453** (2003), pp. 385–405
16. I.C. Sluimer, P.F. van Waes, M.A. Viergever, B. van Ginneken, Med. Phys. **30**, 3081–3090 (2003)
17. N. Lee et al., Am. J. Roentgenol. **501** (1997), pp. 234–240
18. S. Xie, S. Shan, X. Chen, J. Chen, IEEE Trans. Image Process. **19**(5), 1349–1361 (2010)

Economic Sustainability of Integrated Hydropower Development of Ero-Omola Falls, Kwara State, Nigeria

Lawal Kola Maroof, Bolaji Fatai Sule, and Ogunlela Ayodele Ogunlela

Abstract Economic sustainability of integrated hydropower development of Ero-Omola Fall was investigated in compliance with Hydropower Sustainability Development Protocol (HSDP) developed by International Hydropower Association (IHA) in 2004. Field work was carried out to obtain primary data like streamflows, sediment characteristics, petrographic information, water quality, water quantity demand, hydropower load demand and land topography. Economic optimization of hydropower generating potential of Ero-Omola Fall, integrated with water supply, irrigation and drainage was carried out based on the data obtained from the field work. The potential hydropower of Ero-Omola Water Fall was estimated at 8.0 MW, Water supply to communities is estimated at 18 Mm^3/day, irrigation water for Fadama farmers at 2.2×106 m^3 and ecological water release of 0.0504×106 m^3 were also derivable from the scheme. The modified internal rate of return for hydropower, water supply and irrigation yielded the highest returns of 13 % on capital, while hydropower alone yielded 5 %. The NPV of cumulative generated cash flows is positive, which indicates that the project would not operate at a loss. The findings also showed that, the sustainable conjunctive use of hydropower releases is the most sustainable mitigation measures against seasonal flooding downstream of the proposed hydropower plant. The study has established a rational basis for the assessment of a typical medium scale hydropower plant which could be adopted for similar locations in Nigeria.

Keywords Water quantity demand • Hydropower load demand • Land topography

L.K. Maroof (✉)
Nigeria Integrated Water Resources Management Commission, Abuja, Nigeria
e-mail: maroofkular@gmail.com

B.F. Sule
National Centre for Hydropower Research and Development, University of Ilorin, Ilorin, Nigeria

O.A. Ogunlela
Agricultural and Bio-Systems Engineering Department, University of Ilorin, Ilorin, Nigeria

A.M. Gil-Lafuente and C. Zopounidis (eds.), *Decision Making and Knowledge Decision Support Systems*, Lecture Notes in Economics and Mathematical Systems 675, DOI 10.1007/978-3-319-03907-7_15, © Springer International Publishing Switzerland 2015

1 Introduction

There are several thermal power plants and hydropower plants in Nigeria with total installed capacity of 8,664 MW [1]. The combined installed capacity of the three major hydropower stations in Nigeria (Kainji, Jebba and Shiroro) is estimated at 1,900 MW, The estimated demand is 10,000 MW while the available capacity was 5,514 MW in 2012 [1]. The situation is compounded by the failure of the existing power stations to replace vital spare parts due to unsustainable cost recovery mechanism as well as downstream flooding and inundation of farmlands between September and October which has become an annual phenomenon in Nigeria. The inability of the hydropower stations to operate at installed capacity could be attributed to many reasons amongst which are [2];

(a) Hydrological factors, such as (1) seasonal variation in flow to the reservoirs (2) inter-annual variation in flow to the reservoir (3) conflict among competitive uses and (4) sedimentation.
(b) Non-hydrological factors, such as (1) maintenance and spare part problems (2) inadequacy of funds (3) human resources and (4) engineering economics.

In order to solve these problems many authors [2–6] have carried out reservoir optimization studies to model hydropower releases for optimum benefits. Even though the benefit of optimal hydropower reservoir policy is to reduce seasonal flooding of downstream plains, it does not however optimize the usage of continuous hydropower releases which could as well provide sustainable potable water supply, irrigation and flood control for the benefits of downstream communities. It is widely believed that reservoir operations policy alone may not guarantee security against seasonal flooding. The formulation of sustainable conjunctive use of hydropower releases is the best mitigation measures against seasonal flooding of farmland. Conjunctive use of hydropower releases involves provision of fish passes, water supply facility, irrigation and drainages as well as ecological water balance for downstream eco-systems [7]. It has also proved to be the most effective and most sustainable ways of controlling flood since almost 90 % of releases would be diverted for useful purposes. The conjunctive use of hydropower releases also ensures that economic activities of benefitting communities are not disconnected by developmental projects. A sustainable water resource system is one designed and managed to fully contribute to the aspiration and desires of the benefiting communities, now and in the future, while maintaining their ecological, environmental and hydrological integrity [7]. Economic efficiency and fiscal sustainability demand that the capital costs of hydropower, water supply and irrigation infrastructure should be recovered from the users in order to permit longer-term replication of investments. For an investment to be worthwhile, the expected return on capital must be greater than the cost of capital. The cost of capital is the rate of return that capital could be expected to earn in an alternative investment of equivalent risk. Sustainability development protocol of a hydropower project requires that the following activities are carried out accurately and evaluated before development is contemplated:

(a) An accurate prediction of inflow to hydropower plant is critical to sustainable overall prediction of energy derivable from such sites.
(b) Raw water quality must be established to determine its impact on mechanical components of the project and to protect it from corrosion activity as well as to develop an effective treatment plant.
(c) Sediment characteristics of potential site must be carefully determined, so as to ensure that turbines runner is protected to last longer.
(d) Drainage facilities must be provided to divert unwanted flood water from hydropower facilities.
(e) Appropriate pricing mechanisms must be developed based on affordability and willingness to pay.

It is becoming increasingly recognized that poor performance of hydropower project is not only a consequence of poor management problem alone, but that many of the problems stem from inadequate economic justification ab initio. This situation could be attributed to a number of reasons as identified by [8, 9] include:

1. I inordinate focus on project design and construction.
2. Inadequate consideration of routine operation and maintenance issues once the project is completed.
3. New unplanned issues which may arise, but were not originally considered.
4. Conflict and competition among competing uses during drought period.
5. Complex legal agreements, regional issues and pressure from various special interests.

Thus, attention must focus on improving the economic sustainability, use of effective price mechanism and efficiency of reservoir operation to maximise the benefits of such projects and to minimise adverse effect on the environment.

The Federal Government of Nigeria initiated an electricity reform process from 2001 to 2005. The new reform among other things unbundled the Power Holding Company of Nigeria (PHCN) and brought about the establishment of Nigeria Electricity Regulatory Commission (NERC) in 2005. The new reform allows for provision of electricity by private investors. The purpose is to inject both private and foreign investment into the power sector and allow for appropriate pricing. It is hoped that appropriate pricing will guaranty:

(a) Recovery of an appropriate return on capital invested, depreciation and replacement of capital and recovery of operation and maintenance including overhead cost.
(b) Appropriate electricity tariff as the key to cost recovery and underpins the long term viability of power projects. Currently prices charged do not reflect the true cost of providing electricity services in Nigeria.

(c) Achievement of an economical, sustainable and efficient allocation of resources in a free market economy where producers and consumers would be paid and pay, respectively, for costs associated with services so produced.

The reform also provides a special hydropower intervention fund at the Bank of Industry (BOI), United Nation Development Programme (UNDP) and Central Bank of Nigeria at 5 % interest rate, so as to boost electricity through Public-Private-Partnership (PPP), Nigeria Independent Power Producers and the National Integrated Power projects (NIPP) arrangement.

This study seeks to analyze capital investment and operating costs of an integrated hydropower scheme at the Ero-Omola Falls over the useful life of the project with a life-cycle assessment of alternatives forming an integral component of evaluation processes. The Ero-Omola Fall is located along Osi-Isolo-Ajuba Road off Osi-Idofin road in Oke-Ero LGA, Kwara State, Nigeria. It is about 116 km from Ilorin, the capital of Kwara State. The height of the fall is about 59.01 m. The catchment area of Ero-Omola-Falls is about 145 km^2 with contribution from two rivers namely, Ero-river from Iddo-Faboro near Ifaki in Ekiti State and Odo-Otun river from Ajuba. Ero-Omola Falls lies between latitude 08° 09′ 34.6″N and 08° 09′ 30.8″N and between longitude 05° 14′ 07.4″E and 05° 14′ 06.7″E. Figure 1, shows map of Kwara state and the location of the falls near Ajuba village.

2 Study Approach and Methodology

2.1 Stream Gauging, Discharge Measurement and Rating Curve

Various site visits were undertaken to facilitate gauge installation and hydraulic head survey. Gauge readers were effectively engaged by June 2009 and have since continued to monitor the gauge till date. A staff gauge is the simplest device for measuring river stage or water surface elevation. The staff gauge is a graduated self-illuminated strip of metal marked in metres and fractions thereof. Water levels were read daily, recorded and collated on monthly basis. Limited numbers of discharge measurements (10 Nos.) were undertaken each month for a range of stage to define a relationship between stage and discharge at the two gauging stations. Discharge measurements taken at various times were used to generate the discharge rating curves and to establish the minimum and maximum water levels.

The stage-discharge relation, which is the rating curve, is then combined with continuous periodic stage measurements to record discharge as well as stage simultaneously. The rating curve was converted to discharge. In general for a gauge height $H(m)$; the discharge $Q(\text{m}^3/\text{s})$ is related to height $H(m)$ as [10]:

Fig. 1 Project location map

$$Q = K(H + / - H_0)^n. \tag{1}$$

When $H_o = 0$.
The rating equations is given as [19]

$$Q = KH^n. \tag{2}$$

Where:
Q = Discharge (m³/s)
H = Gauge Height (m)
H_o = Gauge Height when the flow is zero (m)
n and k are constants
This is a parabolic equation which plots as a straight line on double logarithmic graph sheet. K & n are determined using the least square methods. Taking logarithms of both sides of Eq. (2) we obtain the relation:

$$\log_a(Q) = \log_a(K) + n\log_a(H). \tag{3}$$

Which is of the form $y = a_0 + a_1 x$, where $y = \log Q$, $a_0 = \log K$, $a_1 = n$; $x = \log H$.
Then k and n can be calculated from the formulae $a_0 = \log K$ and $n = a_1$.
Taken summation on both sides and assuming N pairs of observation gives, then:

$$\sum \text{Log}_a(Q) = N\text{Log}_a(K) + n\sum \text{Log}_a(H). \tag{4}$$

Multiplying both sides by log H gives:

$$\sum \text{Log}_a(Q)\text{Log}_a(H) = \text{Log}_a(K)\text{Log}_a(H) + n\sum \left(\text{Log}_a(H)\right)^2. \qquad (5)$$

These two equations were solved simultaneously to determine constants k and n and hence rating equation of each month.

2.2 Streamflow Extension

Inadequate hydrological data may lead to over or under design of the power plant. Stochastic theory is applied in order to minimize the risk of such uncertainties. The stochastic theory provides opportunity to forecast and extend short duration data in a planning process [11]. If hydro-power projects are planned and designed on the basis of rather short time series of observed hydrological data the danger of inaccurate solutions is high. When only short term data are available at project site, the short term data is normally extended with the help of long term data of other sites on the same stream or in the adjoining catchments [12]. The Thomas Fierring method [13] was employed to extend the 12 months data obtained at Ero-Omola Falls.

2.3 Demographic Data

Population is a major driver of energy demand. Other important determinants of energy demand include the level of economic activity and its structure, measured by the Gross Domestic Product (GDP). From the demographic data, the projected population was used in the estimation of energy demand of the benefitting communities. The project catchment areas comprises of three local government areas namely: Ekiti, Oke-ero and Isin, LGAs with a combined population of 172,207 [14]. This is projected to 2036 using the average national population growth rate of 2.83 %.

2.4 Meteorological Data

Meteorological data such as temperature, wind speed, sunshine hour, relative humidity and rainfall were collected at Omuaran, Kwara state for the purpose of estimating the evaporation losses. Meteorological variables were also collected from ECWA Primary School at Osi, Kwara state for the estimation of the crop water requirement for vegetables, maize and sugarcane which are the crops planted downstream by Fadama farmers. The crop coefficients were obtained from International Institute for Tropical Agriculture in Ibadan, Nigeria.

2.5 Determination of Sediment Characteristics and Mineral Composition

Scientific evaluation of sediment inflow, its distribution in sizes and gradation are essential for sustainable management of hydropower project, both in the long and short term. The gradation of sediment is paramount in the selection of suitable turbine blades and vanes. The sediment samples from Ero-river were analyzed at the University of Ilorin Civil Engineering Department to determine the sediment load. Sediment can cause damage and sometimes very serious damage to under water components of the generating equipment such as runners, guide vanes, etc. resulting in loss of power generation and costly repair and maintenance of equipment. It has been observed that high concentration of even fine angular quartz particles can cause maximum erosion in most hydropower plants. A variety of sediment exclusion and extraction measures must be provided to reduce size and concentration of sediment particles in the flow reaching the generating equipment in order to reduce damage to a power plant. The planning and design of these measures depend on the sediment characteristics. Hence the sediment characteristics like, size, shape, hardness and concentration which are site specific must be assessed with as much accuracy as possible for planning and design of cost effective sediment exclusion and extraction measures at intake [15].

Petrographic analysis is the determination of mineral composition of the sediment. Samples taken from sites were analyzed in the Geology Laboratory at the University of Ilorin. As the plant components coming into contact with the water such as slide valves, pressure pipes, turbines, rotors and casings are destroyed by the suspended matter, it is necessary to determine its composition and concentrations. One of the essential requirements for the design of a hydropower plant is that the water drawn in should be free of sediment as far as possible. The presence of sediment, especially sharp-edged sand particles, may cause wear of the turbine runner vanes and other steel parts besides causing damage to the tunnel lining. Abrasion effects become more pronounced with increase in head.

2.6 Raw Water Quality

In order to effectively control aggressive corrosion, the chemical analysis of water is important to have knowledge of presence of salts and the nature of water (acidic or alkaline) which could affects metals, equipment and concrete structure. The raw water was also tested for physical and chemical characteristics to determine the type of treatment required for suitable water supply. The physical, chemical and microbiological parameters tested in the laboratory were compared to the permissible limits set by Nigeria Water Quality Standard [16].

2.7   Survey Works

Preliminary survey work was carried out to develop topographical maps, hydraulic head, choice of dam axis, pipeline route for the transmission of raw water and total length of required penstock. The gross head is the vertical distance that the water falls through in giving up its potential energy (i.e. between the upper and lower water surface levels). Having established the gross head available, it would be possible to estimate losses, from trash racks, pipe friction, bends and valves. The hydraulic head profile along the Ero-Omola river course with the gross head were estimated from the topographic survey mapping of the sites. The gross head minus the sum of all the losses equals the net head which is available to drive the turbine.

2.8   Electrical Load Demand Survey and Load Projection

Five households each from the three LGAs within the project catchment areas and four other adjoining LGAs were surveyed for 30 days. A template or checklist for determination of historical load profile within the LGAs was developed and was distributed and monitored for 30 days. The survey was undertaken with the assistance of Zonal office of PHCN at Omu-aran, Kwara State. The peak daily demand load and future typical load forecast were attained with the model equation developed by International Atomic Energy Agency (IAEA) known as the Model for Analysis of Energy Demand (MAED). The demand projection method of MAED is expressed by the following equation [17].

$$E_i = e_i * VD_i = e_0 * F_i * VD_i. \tag{6}$$

Where E_i = energy demand in year i for (i = 2010–2036); e_0 = base year energy intensity; e_i = the value in year i; VD_i = value of the determinant of energy demand in year i; F_i = modifier of e_0 for year i; it depends on such factors as penetration of technology, energy use efficiency, economy, life style, demography, and other social-economic and technological factors that affect energy demand, in year i relative to the base year.

2.9   Assessment of Hydropower Potential

The following are the basic steps used to compute average annual power based on the guidelines of the [15]:

1. Determination of the flow losses: Flow losses such as consumptive losses include reservoir surface evaporation losses and diversion such as for irrigation and water supply. Non-consumptive losses include ecological requirement,

leakage through/around dams and embankment structures and leakage around spillway or regulating outlet gates.

2. Development of head data—A head versus discharge curve to reflect the variation at tail water elevation with discharge.
3. Selection of plant site: First the plant hydraulic maximum discharge that can be met through the turbine is selected. For preliminary studies the initial plant size was based on the average annual flow or a point between 15 and 30 % exceedence on the flow duration curve. Next the net head corresponding to the assumed hydraulic capacity was identified.
4. Definition of usable flow range and derivation of head–duration curve: The flow duration is reduced to include only the usable flow range, because the turbine characteristics limit the stream flow that can be used for power generation using the flow–duration data and time versus discharge data.
5. Derivation of Power Duration Curve: the energy available at 100 % exceedence is derived from power versus flow time.

2.10 Engineering Economics and Financial Analysis

Financial feasibility is the evaluation of the ability of the project to provide debt service from the capital required to construct and operate the project. Economic feasibility is the evaluation of project costs and benefits with the project deemed feasible when benefits exceed costs. In this study, cash flow represents all quantified costs and benefits, so that the financial analysis provides the cost (disbursement) and benefits (receipts) for the economic analysis. The economic criterion used is the Financial Internal Rate of Return (FIRR). The economic evaluation of hydropower development plans combines basic methods of engineering economics with benefit estimation procedures. Analyses of economic costs and benefits provide important information for use, along with various other forms of information, in making a myriad of decisions in planning, design, operations, and other water resources engineering activities [18].

3 Design of the Integrated System

3.1 Layout of the Scheme

Figures 2 and 3 show the hydraulic head profile along the Ero-Omola river course with the gross head estimated at 59.01 m. The gross head minus the sum of all the losses equals the net head, which is available to drive the turbine. The longitudinal profile of the proposed dam axis is shown in Fig. 4, while Fig. 5 shows the layout of the integrated system. The design of each component was carried out using all the data and information described in Sect. 2 above.

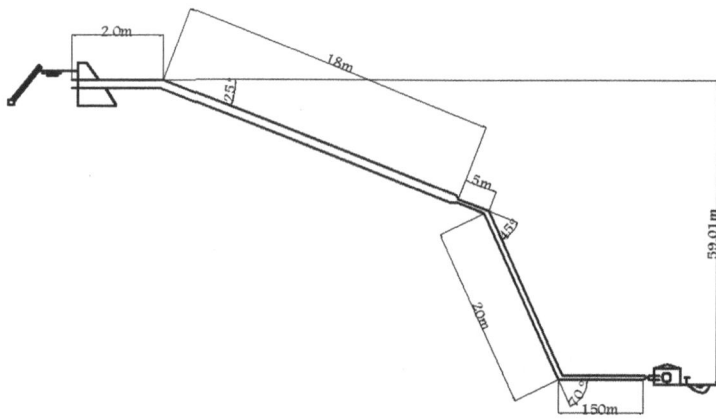

Fig. 2 Hydraulic head survey

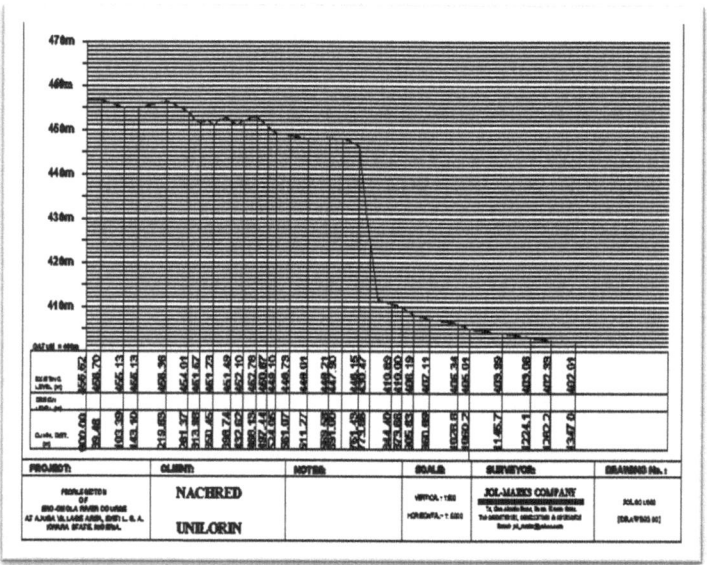

Fig. 3 Hydraulic head profile

3.2 Hydropower System Components

The component of hydropower system consists of intake channel, trash rack, sediment tank, forebay tank, overflow conduit, penstocks, draft tube, tailrace channel, power house, turbine and generator. The designs of these components were carried out to allow for costing of the project. The summary of the main features of the components are;

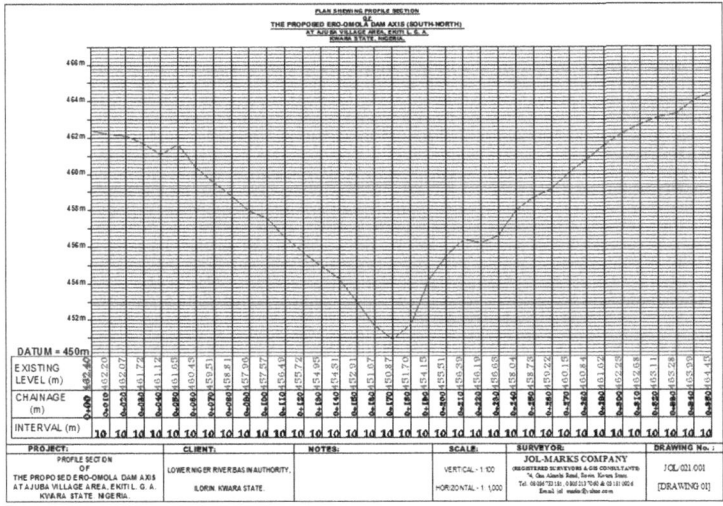

Fig. 4 Longitudinal profile of dam axis

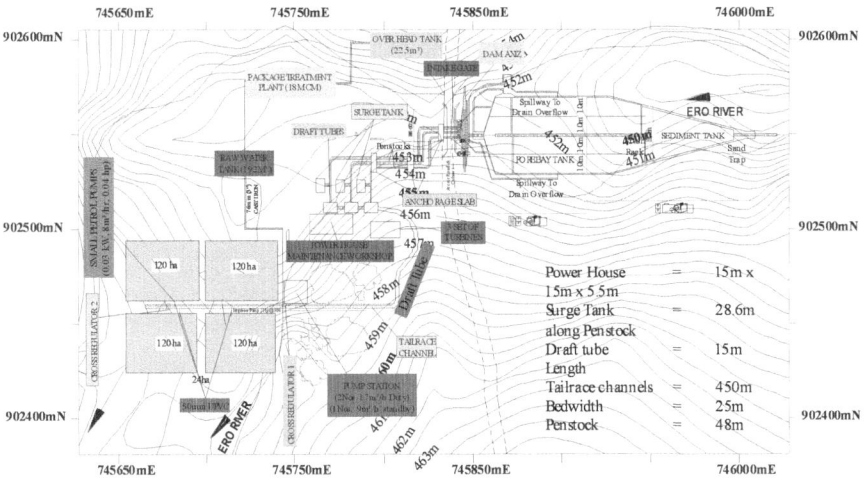

Fig. 5 General layout of integrated hydropower scheme

1. Intake Channels.

   The width of the river at intake site is 24 m. It is proposed to construct a simple stone masonry channels to divert water into the sediment tank. The length of intake channel from intake to sediment tank is about 600 m. The parameters of the channel are given below:

   - Type = Trapezoidal.
   - Discharge, $Q = 24.01$ m^3/s.

- Bed slope, $S = 1/600$.
- Rugosity Coefficient, $n = 0.016$.
- Water depth $= 3.5$ m.
- Bed width $= 24$ m.
- Free board $= 1.50$ m.

2. Trash Rack

   At the intake of channel, a trash rack and steel gate is provided to check the floating material and to control the flow. Trash rack is to be fabricated with Mild Steel flats with 25 mm clear spacing centre to centre. The parameters of the Trash rack are given below:

   - Mild steel mesh $= 25$ mm c/c.
   - Velocity $= 3$ m/s.

3. Design Head and Discharge.

   The following levels were measured between the power house (ground floor) and permanent bench mark position:

   - Water level at intake $= +462.40$ m.
   - Water level at forebay $= +461.02$ m.
   - Water level at tailrace $= +402.01$ m.
   - Gross head $= 461.02 - 402.01 = 59.01$ m.
   - Head loss $= 462.40 - 461.02 = 1.38$ m.
   - Gross Flood $= 24.01$ m^3/s.
   - Design flood $= 21.00$ m^3/s (80 years return period).

4. Petrographic Information.

   - Grain Size Distribution $= 0.02$–0.2 mm.
   - Quartz Sand $= 0.05$ mm.

5. Sediment Tank Capacity.

   - Length $= 40$ m.
   - Width $= 24$ m.
   - Height $= 24$ m.
   - Freeboard $= 1.5$ m.
   - Slope angle $= 26.60$.
   - Bed slope $= 1:600$.

6. Forebay Tank.

   The forebay tank is located on a relatively flat area followed by the penstock provided along moderately slopping side leading to power house on a flat terrace. A 250 mm dia pipe is also provided in the forebay for flushing of silt. The length of flushing pipe is about 180 m.

   - Design discharge $= 7.00$ m^3/s.
   - Storage required $= 180$ min.
   - Water depth $= 24.0$ m.

- Width = 24.0 m.
- Length = 40.0 m.
- Free board = 1.50 m.

7. Overflow Conduit.

- Type = HPDE.
- Discharging Capacity = 7 m³/s.
- Length = 180 m.
- Width = 4.5 m.
- Depth = 3.5 m.
- Material = Stone masonry.
- Slope = 1:60.
- Velocity = 3 m³/s.

8. Penstock.
The design features of penstock are given below.

- Penstock material = mild steel.
- Design discharge = 7.00 m³/s.
- Length = 48 m.
- Diameter = 1,200 mm.
- Velocity (say) = 3.0 m/s.
- Area of pipe require = 7.00/3.0 = 2.3 m².
- Provide, diameter of pipe = 1,200 mm.
- Thickness of Pipe = 16 mm.
- Bends = 4 No.
- Expansion Joints = 4 No.

9. Head loss.

- System Friction loss = 2.5 m.
- Anchor Blocks = 150 m c/c.
- Surge Tank = 28.6 m along penstock.
- Area of inlet of surge tank = 180.83 mm².

10. Draft Tubes.

- Material = mild steel.
- Diameter = 1,200 mm.
- Length = 15 m.

11. Tail Race Channel.
Tail race channel is designed as trapezoidal section to handle design discharge of 21.00 m³/s. The length of tail race channel is about 240 m.

- Design Discharge = 21 m³/s.
- Length = 240 m.
- Bed Slope, S = 1/500.
- Roughness Coefficient, n = 0.016.

- Water depth, d = 6.0 m.
- Bed width, B = 25 m.
- Free board = 2.5 m.

12. Power House Building.

The layout of the power house has been worked out on the basis of use of standard Cross flow turbine The dimensions of the power house are 15.0 m $(L) \times 15.0$ m. $(B) \times 5.5$ m (H) other details are:

- Type = Surface Power House.
- Installed Capacity = 3 unit of 2,500 kW.

13. Turbine.

- Type = Francis Cross Flow Reversible pump.
- Number = 3 units.
- Rating = 2,500 kW.

14. Type of Generator.

- Type = Synchronous.
- Nos. = 3 units.
- Capacity = 65 KVA with 25 % continuous over load capacity, 0.415 kV, 3 Phase, 0.9 pf, 50 HZ, 3 %, 2,500 rpm, Class "F" Insulation.

3.3 Water Supply

Water demand estimates based on population of the three LGAs, as presented in Table 1:

The daily water requirement for the three LGAs is estimated at 46,495.89 m^3 or approximately 46,500 m^3/day.

Water Quality Design Principle is based on the Nigeria Standard for Drinking Water Quality (NIS 554:2007). Hence the treatment system comprises of coagulation, flocculation, sedimentation, filtration and chlorination. Others are:

- Daily Water Demand = 46,500 m^3/day.
- Static Head = 506.857 − 450.939 = 55.918 m.
- Pump capacity = 32 m^3/s High lift.
- Treatment plant capacity = 18 Mm^3/day.
- Elevated (Clear Water Tank) = 22,500 l (22.5 m^3) or 5,000 Gallon.

Table 1 Water demand estimates

Category	Population	Litres/day	m^3/day
Population			
(i) 3 LGAs	172,207 (C)	©150	25831.05
Domestic demand			
(ii) Residents 80 % of 172,207		137,765.6	20,664.84
(iii) Non-Resident (i–ii)		34,441.40	5,166.21
(iv) Visitors and services 5 % of C		8,610.35	1,291.55
(v) Industrial Demand 30 % of C		51,662.1	7,749.31
(vi) Institutional Demand 20 % of C		34,441.4	5,166.21
(vii) Fire fighting 10 % of C			2,583.10
(viii) Leakages, Houses waste unaccounted water 15 % of C			3,874.65
Grand Total			46,495.89

3.4 Irrigation Water Requirements

The computed results for the Fadama irrigation scheme are peak water requirement and total gross water requirement. Irrigation crop water requirements were estimated at 2.2×10^6 m^3 using the FAO CROPWAT 8.0 software, irrigation and drainage parameters are presented below:

- Number of Fadama farmers = 56.
- Climatic data = 35 years.
- Peak Irrigation Water Requirement = 0.43 l/s/ha.
- Gross Irrigation Water Requirement = 7.568×106 m^3.
- Drains type: Reinforce Concrete Rectangular.
- Length of main drain = 3,627 m.
- Capacity = 2 m $\times$ 1 m $\times$ 1 m.
- Length of feeder drain = 1,373 m.
- Capacity = 1 m $\times$ 1 m $\times$ 1 m.

4 Financial and Economic Analysis

4.1 Capital Cost Element of Hydropower

The financial and economic evaluation of the project has been prepared according to the Hydropower Sustainability Development Guidelines based on a project life of 30 years. The capital costs for each component were estimated as shown in Table 2. The cost elements for hydropower were based on [7]. The breakdown is shown in Table 3.

Table 2 Project cost summary

S. no	Description	Cost millions (N)
1.	Hydropower	750
2.	Water Supply	245
3.	Irrigation and Drainage	108
	Total	1.103

Table 3 Capital cost element of hydropower

Capital cost element of hydropower	% of total cost
Planning and Design/Supervision	3–10
Civil works	15–45
Mechanical/Electrical	25–55
Electricity Distribution	8–12
Interest during Construction	5–10
Contingency	5–10
Total Capital Cost	
Running Cost	
Fixed cost	5–10
Variable Cost (O&M)	2–5
Contingency	2–5

Source: [8]

The total financial cost of development of Ero-Omola Fall is estimated at N1,103,000,000.00 which include the headworks, civil, electrical and the mechanical components.

4.2 Amortization Analysis

The discount factor computed at 5 % interest rate is [7]:

$$A = F\left(\frac{i}{(1+i)^N - 1}\right) = A = F\left(\frac{0.05}{(1+0.05)^{30} - 1}\right) = 0.01505. \qquad (7)$$

Amortization of various components of the project costs outlined above in Table 2, was computed at an interest rate of 5 %. The total sum of N16,538,300.00 annually over the assets life of the project (30 years) is expected to be deducted installmentally in order to pay back interest and the capital. The projected cash flow is presented in Fig. 6. The cash flow indicates that Ero-Omola project is expected to grow by 5 % profit beginning from year 2015. The total Kilo Watt Hour unit available annually at 90 % dependability is 15,137.28 MWh.

HYDROPOWER,WATER SUPPLY AND IRRIGATION	CONSTRUCTION PERIOD		2013	2014	2015	2016
	2011	2012				
Hydropower Development						
Installed Capacity(kw)	8000	8000	8000	8000	8000	8000
Annual Electricity Delivered to Grid						
(10³kwh)	0	0	7008	14016	14016	14016
Tariff(N14.00/kwh)	0	0	14	14	14	14
Cash Inflows	0	0	98.112	196.224	196.224	196.224
Annual Electricity Sales(10⁶) Naira	0	0	98.112	196.224	196.224	196.224
Municipal Water Supply						
Installed Capacity (MCM)	18	18	18	18	18	18
Water Delivered to Communities (m³/day))	0	0	23,250	46,500	46,500	46,500
Water Tariff(N45/m³ /day)	0	0	15	15	15	15
Water Supply Revenue/Annum(10⁶) Naira	0	0	127	255	255	255
Irrigation Water Supply						
Installed Capacity (mcm)	2	2	2	2	2	2
water Delivered to farmers(m³/day)	0	0	22	43	43	43
Irrigation Tariff(N21/m³/day)	0	0	21	21	21	21
Irrigation Revenue/Season (Naira)	0	0	82,625	165,518	165,518	165,518
TotalCcash Inflow (10⁶) Naira			225	451	451	451
Total Cash outflows(10⁶)Naira	551.5	1103	184.535419	38.1708509	38.1708509	38.1708509
Fixed Assets Investment	551.5	1103				
Working Capital(15%)	0	0	165.45			
Operation/maintenance Cost(0.75%)millions	0	0	4.136	8.272	8.272	8.272
O&M Treatment Plant(3%)Millions	0	0	3.675	7.35	7.35	7.35
Vat (5%) Millions	0	0	11.2744188	22.5488509	22.5488509	22.5488509
Company Income Tax(30%)	0	0	0	0	0	0
Inflation Rate 12.6%	0	0	27.058605	54.11724216	54.1172422	54.1172422
Depreciation Charge(20%)	0	0	1103	2.206	2.206	2.206
Net Cash flow	-551.6	-1103	41	413	413	413

Fig. 6 (continued)

4.2.1 Costs and Financing: General Data and Assumptions

(a) Exchange Rate

All costs given are in N and, where necessary converted from US Dollars to Naira at an exchange rate of US$1.00 = N161.00.

(b) Reference Date and Investigation Period

The reference date for all FNPV and FIRR calculations is 01.01.2011. It is assumed that the commercial operation date is 01.01.2013. Thus the first year of operation is a full calendar year with 365 days of operation. The analysis period comprises the years 2013–2043. The planning, tendering and construction period have been assumed to commence by 01.01.2011 and end on 31.12.2012, (i.e. a construction period of 2 years.) The operation period has been assumed to commence on 01.01.2013 and end on 01.01.2043, covering 30 years. A 50 % capacity utilization is assumed in the first year of operation.

(c) Installed Capacity: 8,000 kw.

- Annual electricity delivered to national grid (2013): $8,000 \text{ kw} \times 24 \text{ h} \times 365$ days $\times 0.2$ (flow reduction factor) $= 14,016,000 \text{ kwh} = 14,016$ MWh or 7,008 MWh at 50 % capacity utilization in the first year of operation.
- Electricity tariff = N14.00/kwh.
- Electricity sales = Cash flow = $14,016 \times 10^3$ kwh $\times$ N14.00/kwh = 196.224 Million (2014) or $7,008 \times 10^3$ kwh N14/kwh = N98.112 Million (2013) at 50 % capacity utilization.
- Power delivered to National grid: (2014): 14,016 MWh (100 % capacity utilization).

OPERATION PERIOD

2017	2018	2019	2020	2021	2022	2023	2024	2025	2026	2027	2028	2029	2030
8000	8000	8000	8000	8000	8000	8000	8000	8000	8000	8000	8000	8000	8000
14016	14016	14016	14016	14016	14016	14016	14016	14016	14016	14016	14016	14016	14016
14	14	14	14	14	14	14	14	14	14	14	14	14	14
196.224	196.224	196.224	196.224	196.224	196.224	196.224	196.224	196.224	196.224	196.224	196.224	196.224	196.224
196.224	196.224	196.224	196.224	196.224	196.224	196.224	196.224	196.224	196.224	196.224	196.224	196.224	196.224
18	18	18	18	18	18	18	18	18	18	18	18	18	18
46,500	46,500	46,500	46,500	46,500	46,500	46,500	46,500	46,500	46,500	46,500	46,500	46,500	46,500
15	15	15	15	15	15	15	15	15	15	15	15	15	15
255	255	255	255	255	255	255	255	255	255	255	255	255	255
2	2	2	2	2	2	2	2	2	2	2	2	2	2
43	43	43	43	43	43	43	43	43	43	43	43	43	43
21	21	21	21	21	21	21	21	21	21	21	21	21	21
165,518	165,518	165,518	165,518	165,518	165,518	165,518	165,518	165,518	165,518	165,518	165,518	165,518	165,518
451	451	451	451	451	451	451	451	451	451	451	451	451	451
38.1708509	51.7001614	51.7001614	51.7001614	51.7001614	51.7001614	51.70016144	51.7001614	51.7001614	51.7001614	51.7001614	51.7001614	51.700161	51.70016144
8.272	8.272	8.272	8.272	8.272	8.272	8.272	8.272	8.272	8.272	8.272	8.272	8.272	8.272
7.35	7.35	7.35	7.35	7.35	7.35	7.35	7.35	7.35	7.35	7.35	7.35	7.35	7.35
22.5488509	22.5488509	22.5488509	22.5488509	22.5488509	22.5488509	22.5488509	22.5488509	22.5488509	22.5488509	22.5488509	22.5488509	22.548851	22.5488509
0	13.5293105	13.5293105	13.5293105	13.5293105	13.5293105	13.52931054	13.5293105	13.5293105	13.5293105	13.5293105	13.5293105	13.529311	13.52931054
54.1172422	54.1172422	54.1172422	54.1172422	54.1172422	54.1172422	54.11724216	54.1172422	54.1172422	54.1172422	54.1172422	54.1172422	54.117242	54.11724216
2.206	2.206	2.206	2.206	2.206	2.206	2.206	2.206	2.206	2.206	2.206	2.206	2.206	2.206
413	399	399	399	399	399	399	399	399	399	399	399	399	399

2031	2032	2033	2034	2035	2036	2037	2038	2039	2040	2041	2042	2043
8000	8000	8000	8000	8000	8000	8000	8000	8000	8000	8000	8000	8000
14016	14016	14016	14016	14016	14016	14016	14016	14016	14016	14016	14016	14016
14	14	14	14	14	14	14	14	14	14	14	14	14
196.224	196.224	196.224	196.224	196.224	196.224	196.224	196.224	196.224	196.224	196.224	196.224	196.224
196.224	196.224	196.224	196.224	196.224	196.224	196.224	196.224	196.224	196.224	196.224	196.224	196.224
18	18	18	18	18	18	18	18	18	18	18	18	18
46,500	46,500	46,500	46,500	46,500	46,500	46,500	46,500	46,500	46,500	46,500	46,500	46,500
15	15	15	15	15	15	15	15	15	15	15	15	15
255	255	255	255	255	255	255	255	255	255	255	255	255
2	2	2	2	2	2	2	2	2	2	2	2	2
43	43	43	43	43	43	43	43	43	43	43	43	43
21	21	21	21	21	21	21	21	21	21	21	21	21
165,518	165,518	165,518	165,518	165,518	165,518	165,518	165,518	165,518	165,518	165,518	165,518	165,518
451	451	451	451	451	451	451	451	451	451	451	451	451
51.7001614	51.70016144	51.7001614	51.7001614	51.7001614	51.7001614	51.7001614	51.70016144	51.7001614	51.7001614	51.7001614	51.7001614	51.7001614
8.272	8.272	8.272	8.272	8.272	8.272	8.272	8.272	8.272	8.272	8.272	8.272	8.272
7.35	7.35	7.35	7.35	7.35	7.35	7.35	7.35	7.35	7.35	7.35	7.35	7.35
22.5488509	22.5488509	22.5488509	22.5488509	22.5488509	22.5488509	22.5488509	22.5488509	22.5488509	22.5488509	22.5488509	22.5488509	22.5488509
13.5293105	13.52931054	13.5293105	13.5293105	13.5293105	13.5293105	13.5293105	13.52931054	13.5293105	13.5293105	13.5293105	13.5293105	13.5293105
54.1172422	54.11724216	54.1172422	54.1172422	54.1172422	54.11724216	54.1172422	54.11724216	54.1172422	54.1172422	54.1172422	54.1172422	54.1172422
2.206	2.206	2.206	2.206	2.206	2.206	2.206	2.206	2.206	2.206	2.206	2.206	2.206
399	399	399	399	399	399	399	399	399	399	399	399	399

Fig. 6 Projected cash flow analysis

(d) Municipal Water Supply.

- Installed capacity: 18 mcm.
- Water Delivered to communities $= 46{,}500 \ m^3/day$ (2014) or $23{,}250 \ m^3/day$ @ 50 % capacity utilization in the year 2013.
- Water Tariff $= N15/m^3$.
- Water supply revenue/annum $= 23{,}250 m^3/day \times N15/m^3 \times 365$ days $= N12{,}7293{,}750.00$ or N127.29 Million/Annum (2013) or N255 Million/Annum (2014)

(e) Irrigation Water Requirement.

- Installed capacity = 2.2 mcm.
- Irrigation water delivered to farm = 43.07 m^3/day = peal water requirement Irrigation Tariff = N21/m^3/day.
- No of days per season = 6 months (November-April) = 183 days.
- Irrigation Revenue/Season (2014) = 43.07 m^3/day × N21/m^3/day × 183 = N165,518.00 or N21/m^3/day × 183 = N82,625.00 at 50 % capacity utilization in the year 2013.

4.2.2 Cash Flows

The FIRR represents the level of financial return on the investment and, therefore, the investor's main concern centers around expected cash in-flows. In identifying and projecting cash flows from an income generating project such as hydropower [19]. Income Statement (Profit and Loss Statement) with some justification is commonly employed. It is the compound rate of return that you get from a series of cash flows [20].

Total cash outflows are the total revenue accrued from the sales of electricity, municipal water supply and irrigation water supply. i.e. 2013 = (N98,112 + N127 + N0.082625) × 10^6 Naira = 225 Millions Naira.

Total cash outflows is the total initial construction costs of hydropower, municipal water supply and irrigation water supply provided in the Bill of Engineering Measurement and Evaluation (BEME) i.e. (N750 + N245 + N108) × 10^6 = N1103 × 10^6 Naira (2012). 50 % is assumed to have been utilized in the first year of construction period i.e. N551.6 Million Naira.

Net Cash flow = Total Cash inflow − Total Cash outflow {hydropower operation and maintenance + O/M Treatment Plant + Vat 5 % + Company income tax (after 5 years of Tax holiday i.e. 2018)}, while only the initial working capital is involved in the first year of 2013. Figure 6 shows the cash flow analysis.

4.3 Sensitivity Analysis

Financial analyses were performed with loan periods of 30 years and interest rates of 5, 10 and 21 %. Project cash flow were estimated based on a 30-year project life span. Using these variables, and the energy generation assumptions, The financial performance indicators for the Ero-Omola hydro power plant and configurations were analyzed based on three different scenarios as:

1. First Scenario.
   Loan for the project would be obtained from either Central Bank of Nigeria or the United Nation Development Bank intervention fund with Nigeria Bank of

Industry at an interest rate of 5 %. The project life is assumed to be 30 years. The performance indicators obtained are:

- FNPV = Financial Net Present Value = N4, 286.27.
- FIRR = Financial international Rate of Return = 13 %.
- MIRR= Modified internal rate of return = 8 %.
- WACC= Weighted Average Cost of capital = 7.50 %.

2. Second Scenario.
   Loan for the project would be obtained from Nigeria Bank of Industry direct fund at an interest rate of 10 %. The project life is assumed to be 30 years. The performance indicators obtained are:

- NPV = Financial Net Present Value = N314.79.
- FIRR = Financial international Rate of Return = 13 %.
- MIRR= Modified internal rate of return = 11 %.
- WACC= Weighted Average Cost of capital = 7.50 %.

3. Third Scenario.
   Loan for the project would be obtained from open market or Commercial Bank at an interest rate of 21 %. The project life is assumed to be 30 years. The performance indicators obtained are:

- FNPV = Financial Net Present Value = N294.7.
- FIRR = Financial internal Rate of Return = 13 %.
- MIRR= Modified internal rate of return = 19 %.
- WACC= Weighted Average Cost of capital = 7.50 %.

While the internal rate of return (IRR) assumes the cash flow from a project are reinvested at the (IRR), the modified internal rate of return assumes that positive cash flows only are re-invested at the cost of capital. Therefore MIRR accurately reflect the true cost of viability and profitability of a project than the IRR. The intervention fund with the Bank of Industry in Scenarios 1, appears to be the best option for this project. This fund allows the project to grow at a sustainable level of 7.5 % under the influence of prevailing inflation rate of 12.6 %.

5   Conclusion

Limited research works are available on the economic sustainability of integrated hydropower development in Nigeria. In order to boost energy supply situation in the country, Ero-Omola Fall located in Oke-Ero LGAs of Kwara State has been studied for sustainable integrated hydropower development. The study shows that, several factors like appropriate design techniques, accurate hydrological assessment, sedimentation study, petrographic information's, water quality assessment, as well as engineering economics of integrated hydropower development contribute significantly to economic sustainability of hydropower projects. Economic efficiency and fiscal sustainability demand that the capital costs of hydropower, water supply and irrigation infrastructure should be recovered from the users in order to permit longer-term replication of investments. For a hydropower investment to be worthwhile, the expected return on capital must be greater than the cost of capital or the internal rate of return. The Internal rate of return however, is that discount rate that makes the net present value of a net benefit or cash flow derivable from electricity sales, equal zero or is the maximum interest rate that a project could pay on invested capital, if the project is to recover its investments and operating costs and still break even. It could also be defined as the rate of return on capital outstanding per period while it is invested in the project. The study demonstrate that the formulation of conjunctive use of hydropower releases is the most sustainable mitigation measures against seasonal flooding downstream of hydropower plant. The potential hydropower estimated from this study if developed would reduce electricity generation deficit in the country. The following conclusions are also drawn from the outcome of this study.

1. The potential hydropower generating capacity of Ero-Omola fall at 100 % dependable flow of 80 years return period is estimated at 8.011 MW. The annual average energy is estimated at 14,035.272 MWh and plant efficiency of 0.70 (by the flow duration method).
2. The potential hydropower generating capacity of Ero-Omola fall at 100 % dependable flow of 80 years return period is estimated at 10,091.502 MW. The annual average energy is estimated at 18,401.56501 MWh and plant efficiency of 0.70 (by Simulation).
3. Water treatment plant capacity is estimated at 22,500 l or 22.5 m^3/s or 5,000 gallons/day at N50/m^3 water tariff. Monthly revenue is estimated at N33,750.00/month or N405,000.00.
4. The raw water quality test indicated that the river is safe for purpose for which it is intended.
5. Irrigation water requirement is estimated at 2.2 × 106 m^3 with peak irrigation water demand of 43.07 m^3/day at N21.00/m^3 irrigation water tariff. Monthly revenue is estimated at N27,134.1.00/month or N325,609.00/annum.
6. The engineering economics of integrated hydropower development with water supply, irrigation and drainage facilities yields highest return of 13 % on capital invested at 5 % interest rate.
7. The cumulative generated cash flow is positive for each scenario which indicates that the project will not operate at a loss. The investment is considered worthwhile since the MIRR is higher than the cost of capital. It can therefore be said that the project is economically viable and financially sustainable.
8. Sensitivity analysis under different interest rate regime of 5, 10 and 21 % indicated that intervention fund with Nigeria Bank of Industry offers the best returns on equity.
9. The results show that hydropower project is economically sustainable when it is integrated with water supply and irrigation. It is regenerative when 90 % of the flow is returned to the stream and the ecological water releases is allow to

recharge downstream aquifers. It can therefore be inferred that the project is economically viable and financially sustainable based on the findings outlined above.

10. It is not economically sustainable to develop hydropower plant to stand alone any longer, except where it is practically impossible to integrate other component of the projects. Comprehensive assessment of integrated hydropower potential should be encouraged nationwide.

11. Finally the study shows that the formulation of conjunctive use of hydropower releases is the most sustainable mitigation measure against seasonal flooding downstream of hydropower plant.

References

1. Federal Ministry of Power, *Regional Communication Bulletin*, vol. 17 (2013)
2. O.D. Jimoh, Niger. J. Eng. Ahmadu Bello Univ. Zaria **14**(1), 53–60 (2007)
3. A.W. Salami, A.M. Ayanshola, J. Agr. Res. Dev. (JARD) **8**(11), 149–150 (2006)
4. B.F. Sule, Niger. Eng. **22**(4), 47–53 (1998)
5. B.F. Sule, Water Resour. Manag. **2**, 209–219 (1988)
6. B.F. Sule, Niger. Eng. **27**(3), 10–17 (1992)
7. IHA, Sustainability Guidelines and Draft Compliance Protocol available at http://www.hydro power.org in Man and Biosphere, in *State-of-Knowledge-Workshop Proceedings,* vol. 4, K.L.R.I. (2007)
8. K.M. Labaide, J. Water Resour. Plann. Manag. **118**(1), 71–80 (1993)
9. K.M. Labaide, J. Water Resour. Plann. Manag. **130**(2), 93–111 (2004)
10. C. Punmia, M. Pande, Renew. Energ. World 248–253 (2008)
11. N.C. Matalas, Assessment of synthetic hydrology. Water Resour. Res. **3**(4), 937–945 (1967)
12. V. Warren, E.H. Terence, W.K. John, Water Resour. Res. **22**(9), 1465–1585 (1972). Intent Educational Publishers, New York
13. T.A. McMahon, R.G. Mein, *Reservoir Capacity and Yield* (Elsevier, Amsterdam, 1978)
14. National Population Commission, National Population Census Figures, *National Population Commission Bulletin*, vol. 8, no. 6 (2006)
15. U. S. Army Corps of Engineers, *Engineering Manual*, As-50, 1110-2-4000 (1995)
16. FMWR, Nigeria water quality guidelines. Standard Organization of Nigeria, SON-15 (2007)
17. T.S. Aluko, Potential hydropower assessment of river Niger. *Energy Commission of Nigeria Report*, vol. 9 (2004)
18. J.P. Gittinger, *Economic Analysis of Agricultural Projects*, 2nd edn. (The Johns Hopkins University Press, Baltimore, MD, 1984). Published for Economic Development Institute
19. B. Schwab, P. Lusztig, J. Finance **24**, 507–516 (1989)
20. E. Solomon, J. Bus. **29**, 124–129 (1996)

Statement of the Synthesis Problem of the Intellectual System of Adaptive Management

Vladislav S. Mikhailenko and Mihail S. Solodovnik

Abstract The paper presents the task of creating an intelligent system of adaptive management of the production facility. Neuro-fuzzy network is proposed for the adaptation of the traditional PI-controllers in the regulation of steam temperature. Is proposed f new three-level structure of the intellectual system of adaptive control with predictor of quality of the transition process. Simulation of intelligent system implemented by using neuro-fuzzy network for example boiler steam temperature under parametric perturbations showed successful results.

Keywords Adaptive control system • Intelligent system • Boiler • Superheated steam • Neuro-fuzzy network • PI - regulator

1 Introduction

Imagine the synthesis problem of the intellectual system of adaptive management (ISAM) for dynamic objects, that exposed to uncontrollable disturbances. Suppose that on the technical object of management (TOM) influence the measured disturbance (defining impact) $Z = Z(t)$, external disturbances $N = N(t)$ and control impact $U = U(t)$. Output variables of the object are available for observations $X = X_B(t)$. The dynamics of an object depends on a number of unknown parameters, the set of which is denoted by α. To be a set of possible values of α, that define the class of admissible objects and disturbances. Set a control objective (the transient time), which defines the desired behavior of the TOM.

It is required to synthesize a control algorithm that uses the measured, approximated and extrapolated on the basis of measurements, the experience of experts

V.S. Mikhailenko (✉) • M.S. Solodovnik
Educational and Scientific Institute of Refrigeration, Cryogenic Technologies and
Bioenergetics, Odessa National Academy of Food Technologies, 1/3, Dvoryanskaya St., 65082
Odessa, Ukraine
e-mail: vlad_mihailenko@mail.ru

A.M. Gil-Lafuente and C. Zopounidis (eds.), *Decision Making and Knowledge Decision*   165
Support Systems, Lecture Notes in Economics and Mathematical Systems 675,
DOI 10.1007/978-3-319-03907-7_16, © Springer International Publishing Switzerland 2015

and training values which are independent of $\alpha \in T$, and provides for the achievement of each $\alpha \in T$ set of quality control processes (objective function).

We take the vector of unknown parameters α consisting of the coefficients of differential equations described TOM, as well as factors determining the change of external disturbances. Also, the vector α contains the parameters which describe the immutable and unknown quantities due to the complexity of the description of TOM . Note that for the production facilities, the vector α, as a rule, should be considered as quasi-stationary: the changing is slow (slow dynamics of the processes and the environment due to large thermal inertia of the process control channels and perturbation).

2 The Main Tenets

The scientific problem is a problem of control under uncertainty associated with $\alpha \in T$. Solution of the problem can be represented as a series of stages: first, of the identification process vector α, and then the definition of the control algorithm to provide the desired quality of the management of traditional or intellectual method. Adaptive strategy is implemented on the basis of a three-tier structure (Fig. 1a). Algorithm first level (control algorithm) depends on the parameter vector θ (the parameter vector control), each time $\alpha \in T$ he has to ensure the achievement of the objective function (predictive indicator of the quality of the transition process—the sum of the time to reach the first deviation and the regulation time of the transition process).

$$P = \min_{i=1,n} \left(T_i(G_1) + T_{pi} \right). \tag{1}$$

Where n—the duration of the TOU, the appropriate choice of $\theta = \theta\,(\alpha)$.

Algorithm second level adapts (adjusts) the vector θ so as to ensure the achievement of the objective function of the unknown.

On figure 1: BTP—block of training and prediction, BAI—block of the accumulation of information, SWG—sine-wave generator.

The algorithm of the third level makes measuring the accumulation of information about the behavior of the object and the decisions of the second and third levels, and based on the methods of learning based on qualitative (expert) and quantitative information produces forecasts of OF (the objective function), and to pre-empt the deterioration values OF makes adjustments to the algorithm second level. The algorithm of the third level is the intellectual superstructure over traditional adaptive SAM to achieve optimal values of the OF in the face of uncertainty and unsteadiness. In the initial phase of the intellectual system of adaptive management (Fig. 1b) parameter identification TOM is carried out during start-up and mode works. The identified parameters of TOM are formed as a mathematical model with a transfer function of the reference channel and

Fig. 1 Block diagrams of the proposed ISAM. (**a**) Parametric diagram of ISAM. (**b**) A three-level diagram of ISAM

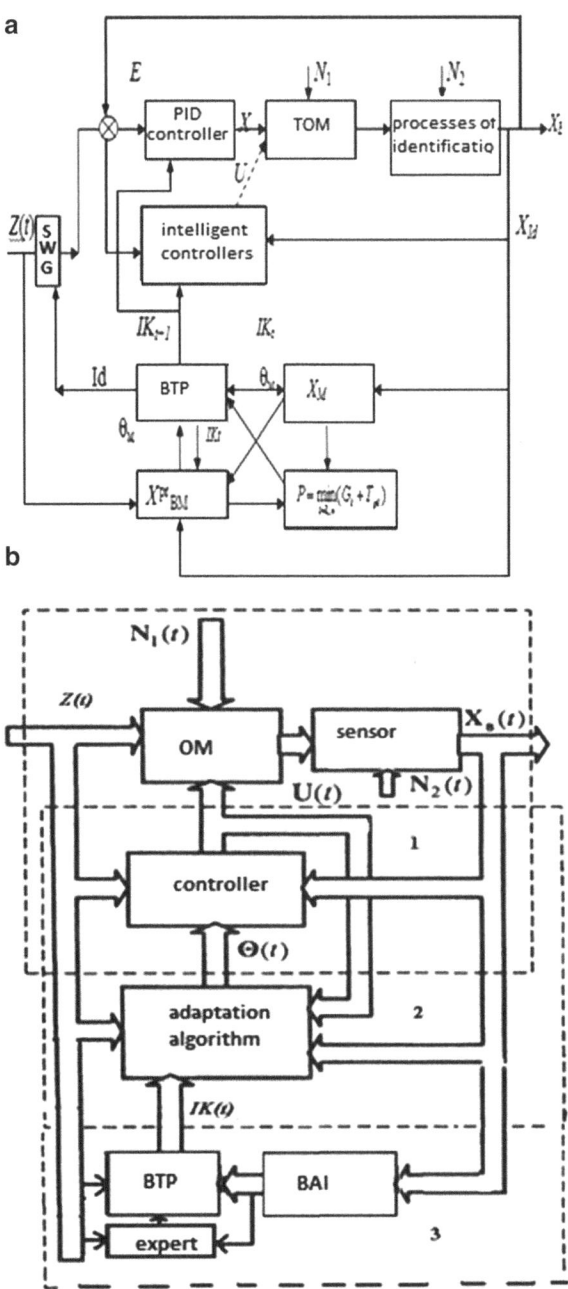

perturbation. And on the basis of their block of training and prediction (BTP) defines adaptive settings for traditional or intelligent controllers. In the course of further work when tasks are changed (load units) to determine the new values of the parameters of the object BTP conducts re-identification by TOM test actions—sine-wave generator (SWG) or a slight change of reference to traditional controller. Object parameters are accumulated in the training set with the help of which take place in the future to forecast the change, and on the basis of predictive mathematical model of BTP are calculated ahead optimal adaptive control action corresponding to the minimum value of the predictive criterion P. Adjustment of the forecast model is made in the BTP by extrapolating the relationship between the parameters and identify the TOM. Later, after the accumulation of sufficient information on the behavior of TOM (dynamic parameters) BTP will be able to determine the impact of adaptive proactive (setting of governors) in the preparation of preliminary information about the expected changes in the job (load) that lead to the minimization of quality transients and conserve resources.

We formalize the task of synthesis. Let the continuous dynamical system is described by a system of equations of state:

$$
\begin{aligned}
\dot{X}(t) &= F\big(X, Z, U, N_1, \alpha, t\big), \\
X_B(t) &= G\big(X, Z, U, N_2, \alpha, t\big).
\end{aligned}
\tag{2}
$$

$F(.)$, $G(.)$—known vector function; N_1, N_2—disturbance on the TOM and noise measurements. $X \in R^n, U \in R^m, Z \in R^r, X_B \in R'$—state vectors, management, external inputs and outputs of the TOM. The purpose of the control is defined as the desired inequality:

$$
q \le \Delta \quad u \quad t \ge t_*, \Delta \ge 0.
\tag{3}
$$

$q(t) = q(X(t), U(t))$—the objective function.

For the problem of tracking as a function of the objective function is selected discrepancy between the actual and the desired trajectory of the object control:

$$
q(t) = q(E(t)), \quad E(t) = X(t) - X_M(t).
\tag{4}
$$

The desired behavior of the system in the traditional adaptive SAM, as a rule, is defined by the reference model:

$$
\dot{X}_M(T) = F_M(X_M, Z, t).
\tag{5}
$$

$X_M \in R^n$—state vector of the reference model, $Z \in R^m$—vector defining influences.

For intelligent adaptive SAM, the behavior of TOM is defined by a self-learning predictive model:

$$
\dot{X}_M^I(T) = F_M\big(X_M^I, O, P, Z, t\big),
\tag{6}
$$

$X_M^l \in R^n$—state vector of the intellectual self-learning model, $O \in R^J$—vector training (test) effects, $P \in R^l$—vector of predicted values, $Z \in R^m$—vector of defining influences.

The problem of synthesis of traditional adaptive SAM (ASAM) is to find a control algorithm of a given class of two-level algorithm of the form:

$$U(t) = U_i(X_B(t), U(t), \Theta(t), Z(t)),$$
$$\Theta(t) = \Theta_i(X_B(t), U(t), \Theta(t), Z(t).$$
(7)

Which provide the achievement of OF in the TOM for each $\alpha \in T$.

Here $U_t(\cdot), \Theta_t(\cdot)$ are some statements. Provided that the adaptive SAM function in a stochastic medium target [Eq. (3)] is replaced by "average goal".

$$Mq \leq \Delta \quad if \quad t \geq t_*.$$
(8)

The problem of synthesis of intellectual SAM is the definition of the algorithm in a given class of three-level algorithm of the form:

$$U(t) = U_i(X_B(t), U(t), \Theta(t), Z(t)),$$
$$\Theta(t) = \Theta_i(X_B(t), U(t), \Theta(t), Z(t),$$
$$IK(t) = IK_i(X_B(t), U(t), Z(t), O(t), P(t)).$$
(9)

$\Theta(t)$—Training, $P(t)$—prognosis.

The system (4), (11) is called the intellectual and adaptive in a class of T in relation to objective function given by Eq. (5) if, for any $\alpha \in T$ and for any initial conditions $X(0), U(0), \Theta(0), IK(0)$ is performed corresponding inequality Eq. (3) or (8).

As part of the task can be considered as time-dependent problems in which the vector of unknown parameters α varies with time ($\alpha = \alpha(t)$).

The vector adaptive controller parameter also depends on the time and corrective action IK. The BTP keeps track of the drift of the unknown parameters, performs their prediction and adaptation to changing conditions. This behavior management is performed in a slow change of α compared to the change of an object state X, in terms of accumulation of quantitative and qualitative information about the drift . In this case, the rapid processes controlled by the first level of the system—control and slow changes are tracked second and third levels . Three levels of management functions in accordance with the division of the rapid movements of the object (coordinate), slow (parametric), probable . At a time when the rate of change comparable to the rate α processes in the facility and the change in external factors , it is advisable to determine the patterns of drift and carry out their prognosis , and the parameters of the law to consider reducing the problem to a new quasi-stationary.

3   Development of the Search Algorithm Intellectual Anticipatory Adaptation with Configurable Model

We establish that the search engine of intellectual identification is measured input and output signals, accompanied by a test signal and effects on an TOM to produce an adaptive mathematical model of the object [1–5]. In contrast to the large number of search algorithms and methods [6–9] Intelligent Identification has the ability of measuring the accumulation of information and expertise on the behavior of the object, self-learning, and can also perform a preliminary forecast value OF at various control and corrective action with the purpose of forestalling deterioration in the quality of transition processes.

Object searching algorithm configuration (Fig. 1b) is a variation of the model parameters θ_M so as to minimize the error criterion function $q\ (E)$.

Let the object $\dot{X}$, object model $\dot{X}_M$, the processes of identification X_{Id} and prediction $X_{\mathrm{BM}}^{\mathrm{pr}}$ are described by equations of state:

$$\dot{X} = f(X, Z, \Theta, t, N_1), X_{Id} = f\left(\psi, A_1, A_2, {}^2T, t\right) + N_2. \tag{10}$$

$$\dot{X}_M = f_M(X_M, Z, \Theta_M, t), X_{BM}^{pr} = f_M(Z, X_B, P, U, IK, In, t_{i+1}). \tag{11}$$

Where $\theta = f(K_{z-u}, T_{z-u}, \tau_{z-u}, n_{z-u}, t_i, K_{N-u}, T_{N-u}, \tau_{N-u}, n_{N-u})$,

$$\Theta_M = f_M\left(K_{z-u}^M, T_{z-u}^M, \tau_{z-u}^M, t_i, K_{N1-u}^M, T_{N-u}^M, \tau_{N1-u}^M, n_{N1-u}^M\right). \tag{12}$$

$$X = f(W_{z-u}(s), W_{N1-u}(s)). \tag{13}$$

$$X_M = f\left(W_{z-u}^M(s), W_{N1-u}^M(s)\right). \tag{14}$$

$$IK = f\left(K_p^A, T_i^A, K_d^A, U_l^A\right). \tag{15}$$

Where In—quantitative and qualitative information on the behavior of the object (training sample predictive controller), U_l^A—adaptive control action for the intelligent controller, K_p^A, T_i^A, K_d^A—adaptive parameters of PID-controller, Ψ—degree of damping, A—amplitude, K_{z-u}—gain control object channel, T_{z-u}—the time constant of the object on the control channel, τ_{z-u}—delay control channel, n_{z-u}—object order over the control channel, $N-u$—channel disturbance, $W_{z-u}^M(s)$—transfer function model, $W_{z-u}(s)$—transfer function object control.

4 Testing of the Intelligent Adaptive ACS for the Example of Superheated Steam Temperature Control Unit

In the domestic power system is mainly used typical system temperature control of superheated steam differentiator [9, 10]. The automatic control system (Fig. 2a) includes a control loop of the PI-controller and an additional measurement of the auxiliary controlled variable formed in the block of differentiation. In actual operation of the system shows that the significant changes in the flow of steam entering the turbine, there is a need to reconfigure the values of the coefficients of its control loop to achieve the desired temperature. In fact, it is an important objective indication that the operation of the control system takes place in the conditions of a priori uncertainty. Analysis of the superheater performance indicates that the control object is a variable transport delay, its dynamic properties substantially depend on the oxygen content in the flue gases, contamination of the heating surfaces and on the regime factors—load type and grade of fuel combusted, conditions of heating surfaces, the excess air, etc. In addition, obtaining a mathematical model of the superheat steam is usually associated with the approximation of the acceleration measured experimentally, and as a consequence of the mathematical description of a priori becomes inaccurate.

Recently in power the traditional methods of active identification and associated algorithms for the optimal settings PI and PID-controller on the analysis of the complex frequency response (CFR) of the objects are spread, as well as methods of self-oscillations [3]. For example, the modified Ziegler-Nichols [4] method used in the Russian adaptive controllers Remicont and Protaras and the SCADA package of Trace Mod. It should be noted that the system temperature steam unstable undesirable process the request process procedure as significant deviation from the normal temperature of the steam can cause premature wear of the turbine and the steam superheater. Thus , there is a scientific problem finding the best methods to identify the object in the case of changes in its load and algorithms for setting the PI-controllers, taking into account the views of an expert technician to ensure the expected transition (aperiodic with minimal regulation).

However, note that the current widespread popularity in the theory of adaptive control is obtained scientific approaches involving the use of neural and neuro-fuzzy networks (hybrid) [11–14]. These systems are successfully implementing the knowledge and experience of experts (fuzzy controllers) as well as have the ability to self-learning (neuro-controllers). Consideration of the technology applied, for example , the identification and adaptation of the superheated steam temperature automatic control system (ACS) is the actual scientific problem.

The aim of the study is the development and training of the hybrid network (HN) to determine the optimal values of settings of the PI-controller (adaptation) in the ACS of superheated steam temperature in the adjustment mode of power.

The structure of the adaptive tuning of the PI-controller is based on the use of active frequency identification methods and identifies the optimal settings of PI-controller by hybrid network.

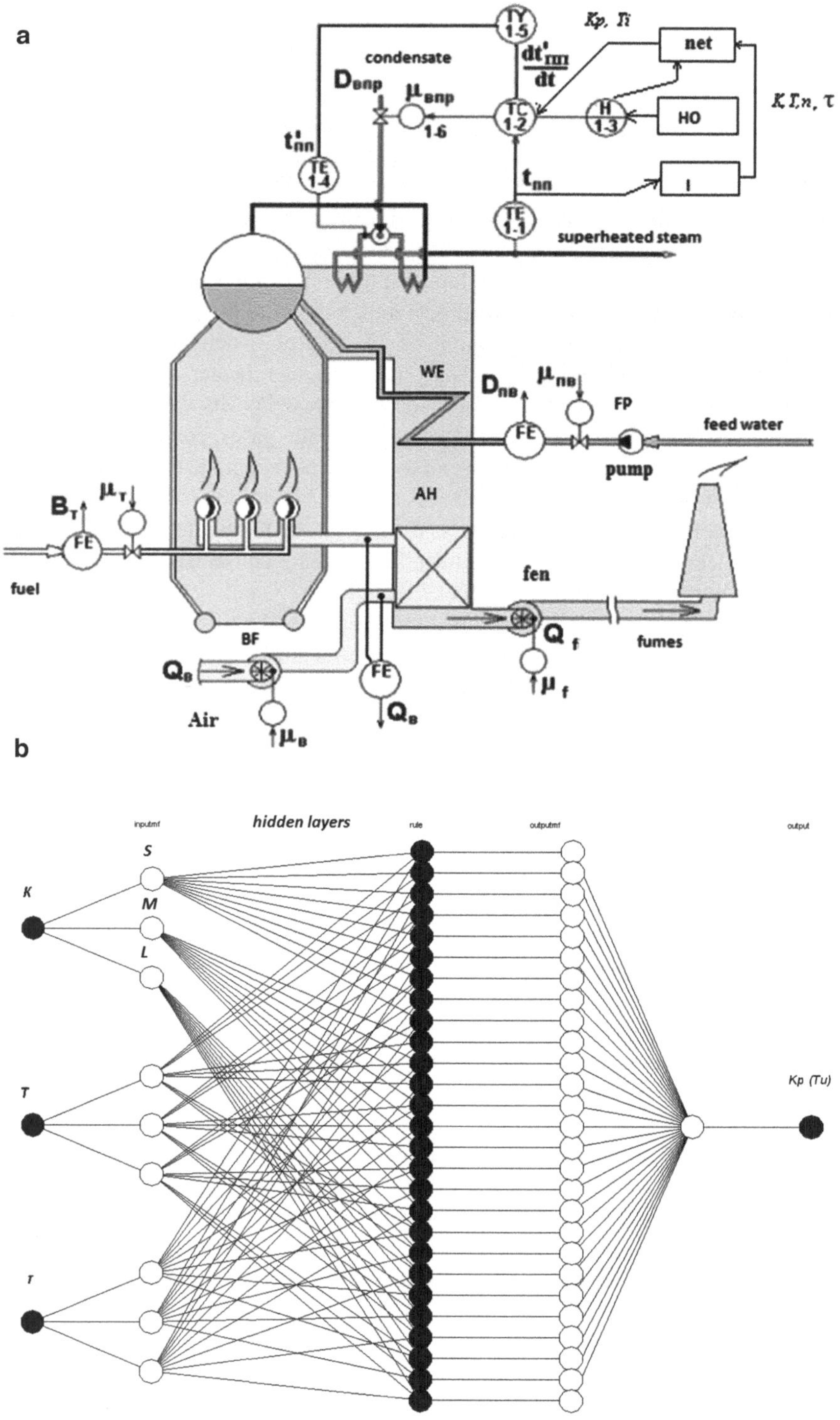

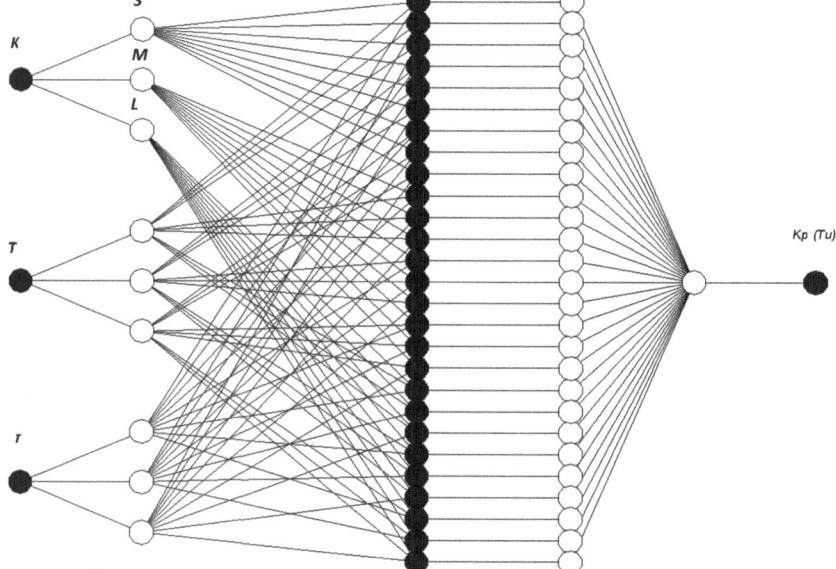

Fig. 2 Superheated steam temperature ISAM with neuro-fuzzy network. (**a**) Functional diagram of superheated steam ACS. (**b**) Structure of the hybrid network

On Fig. 2a: *AH*—Air Heater, *WE*—water economizer, *BF*—blow fan, *FP*—feed pump, *FE*—flow sensor, *TY*—differentiator, *TC*—temperature controller, *H*—a remote control device, $D_{\text{впр}}$—condensate flow, $B_{\text{т}}$—fuel consumption, $Q_{\text{в}}$—air flow, $D_{\text{пв}}$—feedwater flow, μ—the control action, *HO*—harmonic oscillator, *I*—identifier, *FE*—flow sensor. On Fig. 2b: *S, M, L*—fuzzification stage (small, medium, large value of the identified parameters of TOM: gain, time constant, delay), K_{p}, T_{u}—adaptive parameters *PI*-regulator.

The use of frequency methods allowed for immunity algorithm and rational organization of active experiment on the current system in terms of maintaining the stability of the region. The identification was carried out by the complex frequency response (CFR) by applying a system of two sinusoidal input signals from the generator to the differing frequencies belonging to a substantial range. The structure of the transfer function of the object of a powerful steam boiler super heater. Thermal Power Plant via "movement of the regulatory body of cooling water on the desuperheater"—"change the temperature of superheated steam" is composed of several parts of the inertial delay of the form:

$$W(s) = \left(K/(T(s)+1)^n\right) * e^{\tau(s)}.$$

with values which varying within a certain range depending on the operation load or the boiler [2, 14, 15]. The identifies the parameters of the object and its order. In the future, these values are used by the optimizer in the form of neuro-fuzzy network serving the algorithm Takagi-Sugeno [16] to search for the optimum configuration of the PI-controller (K_P, T_i). Education of hybrid network must be performed taking into account the opinions of experts—fixers ACS.

In order to determine the four unknown object parameters (K, T, τ, n), is proposed to use a sine wave generator which produces a complex evaluation of the CFR of the object at two different low frequencies belonging to a substantial range. Given the real equations:

$$\left.\begin{aligned} A_1 &= \frac{K}{\left(\beta^2 \tau^2 \omega^2{}_1 + 1\right)^{0,5n}}; \\ A_1 &= \frac{K}{\left(\beta^2 \tau^2 \omega^2{}_2 + 1\right)^{0,5n}}; \\ \phi_1 &= -arctg(\beta\tau\omega_1) - \tau\omega_1; \\ \phi_2 &= -arctg(\beta\tau\omega_2) - \tau\omega_2 \end{aligned}\right\} \quad (16)$$

Given the real equations whose solution under certain amplitude, frequency and phase shifts: A_1, A_2, ω_1, ω_2, φ_1, φ_2 can find the values of parameters of the transfer function of the object. Time is constant $T = \beta\tau$. It is assumed that the identifier of the experiment the following parameters determined transfer function of the object via regulation:

$$W(s) = \frac{8,27}{(3,05s+1)^3} e^{-0,93}.$$ (17)

Using a technique in the program MathCAD author calculated optimal values of the settings of the PI-controller. It should be noted that calculated by the presented method of tuning of the PI-controller needed for manual adjustment as received a number of damped transient did not meet the specified criteria (first deflection and time management).

For a test sample (learning matrix of neuro-fuzzy network) of the values of optimum parameters of the *PI*-controller, a computer-program experiment in MatLab (Simulink) with expert correction setting values K_p and Ti, the values of the transfer function of the object changed to adapt to different modes of operation of steam boiler.

The program Matlab (ANFIS) [12] was carried out the process of building the adaptive neuro-fuzzy inference (Fig. 2b) for the approximation of the dependence represents a causal link between K, T, τ and K_p, T_i. On the basis of the recommendations [10] and computer simulations in MatLab (Fuzzy Logic Toolbox) was selected amount—EQ. 3 and the type of membership functions (triangular and trapezoidal), describing the input values. The training was used an equal number of cycles 40 and chosen method of training—back propagation.

Thus, the hybrid network was realized that the options of the PI-controller on the characteristics of the control system: $S^k = f(x^k) = f(x_1^k, x_2^k, \ldots, x_n^k)$, $k = 1,2,\ldots N$, if a training set is $((x^1, y^1), \ldots, (x^N, y^N))$.

To model the unknown function f we used an algorithm of Sugeno with the knowledge base of the following type: Π_i: IF x_1 is A_{i1} AND x_2 есть A_{i2} AND x_l is A_{in}, TO $T_н = S_i$, $i = 1,2,\ldots,m$, where A_{ij}—triangular fuzzy sets which describe the expert statements small (S), medium (M), large (L), S_i—the output value of the object. The degree of truth μ, rules i are defined by the conjunction operation. The output of fuzzy system was determined by the center of gravity [12]:

$$T_{ik} = \frac{\sum_{i=1}^{m} \mu_i S_i}{\sum_{i=1}^{m} \mu_i}.$$ (18)

After training the hybrid network and supply to the input parameters of the object ($K_{o6} = 8,27$; $T_{o6} = 3,05$; $\tau = 0.928$) missing in a test sample obtained an experimental and identified as a result of active identification, the network recommended setting values of the *PI*-controller: $T_i = 5,15$; $K_p = 0,12$, and the traditional method of CFR: $K_p = 0,21$, $T_i = 5,649$.

In MatLab (Simulink) an experimental model of ACS with differentiator of the PI-regulators and inertial properties to a third order delay (model superheater via regulation) was developed. In the input of the system was fed a single leap. When

T, 0 C

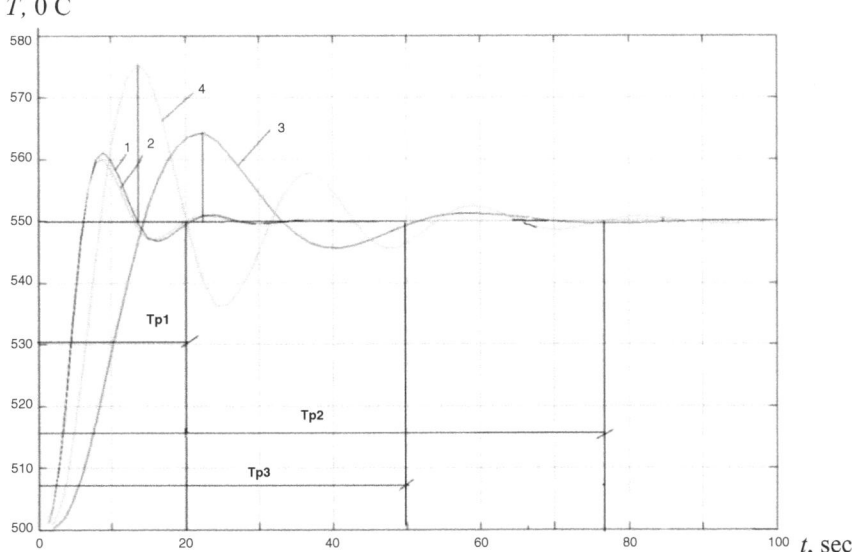

Fig. 3 Transients on channel regulation: 1—traditional ACS (at rated load), 2—neuro-fuzzy SAM (at rated load), 3—neuro-fuzzy ACS (load changes), 4—traditional ACS (load changes)

you change the values of the transfer function of the object in cases of transfer of power to the new regime (simulation steps parametric perturbation) and establishing new parameters object $K_{o6} = 13,7$; $T_{o6} = 5,4$; $\tau = 1.1$, Hybrid System recommended settings $K_p = 0,81$; $T_i = 6,33$, and traditional frequency method for the auxiliary function : $K_p = 0,021$; $T_i = 3,55$. Substitution of these values in the program scheme Simulink provides the following transients (Fig. 3).

5 Conclusion

From the analysis of transients in Fig. 3a (1 and 2) we can conclude that in the nominal or steady operation object, the traditional and hybrid PI-controllers showed identical quality indicators (regulation time $T_{p1} = 20$ seconds (s)), however, under the influence of parametric perturbations (3, 4), the hybrid system has less time to control ($T_{P2} = 50$ s)compared to conventional adaptive ACS ($T_{P3} = 78$ s), also hybrid ACS transition process is aperiodic ($P = 50$ s), traditional $G_{тp} = 50$ % (time to reach $T = 37$ s, $P = 78 + 37 = 115$ s) the degree of attenuation of the hybrid $\Psi_{rp} = 0,91$, traditional $\Psi_{тp} = 0,68$, that is, the proposed hybrid ACS is optimal and energy-saving, and the traditional ACS requires additional adaptation. Also, a comparative analysis of experimental and industrial transients showed that the industrial ACS of coal power plants in Ukraine are self-oscillating, with a long and significant regulatory outputs the values of the steam temperature of range, which indicates the need for the introduction of intelligent adaptive ACS.

Computer program experiments MatLab (Simulink) with varying values of the transfer function of the object (peak simulation mode and starting the boiler), shown in some cases, failure of conventional adaptive approach (the ACS was unstable) unlike algorithm hybrid system transients which were optimal. Based on the obtained results one can conclude that the proposed intellectually—adaptive SAM for example superheated steam temperature has the following advantages over conventional methods and adaptation CFR oscillation currently used systems in power plants:

1. The speed of the process of finding the optimum settings PI-controller with the possibility of approximation and extrapolation of the action and the object of uncertain disturbances.
2. Smaller first deviation and the time of the transition of the regulatory process.
3. The possibility of the optimal functioning of the ACS in all modes of the steam generator.
4. Possibility of using in a process of adaptation of different systems with PI and PID controllers in the thermal power industry.
5. The ability to self-learning, forecasting and accounting knowledge and experience of expert fitters and engineers, as well as working with non-linear objects.
6. Based on the systematic management process of thermal power plants, and claims the author has proposed a structure for intelligent control of thermal power plants, which implements the above methods and algorithms for hybrid technology. Using this structure will significantly reduce production costs, improve efficiency of power, lower fuel consumption, and reduce the influence of psychological factors on the operational staff of power.

References

1. G.D. Krokhin, V.S. Mukhin, I.L. Ivanova, in *IFAC WS ESC'06. Energy Saving Control in Plants and Buildings* (2006), pp. 177–181
2. V.J. Rotach, *Automatic Control Theory* (Moscow Power Engineering Institute (MPEI), Moscow, 2008), p. 396
3. V.J. Rotach, Teploenergetika **8**, 21–26 (1979)
4. V.J. Rotach, *Control Settings Modified by the Ziegler-Nichols* (MPEI, Moscow, 2010), pp. 38–42
5. V.J. Rotach, Teploenergetika **10**, 50–57 (2010)
6. K.J. Astrom, T.T. Hagglund, *Advanced PID Control*, vol. 460 (The Instrumentation, Systems, and Automation Society, Research Triangle Park, NC, 2006)
7. A. S. Kluev, Setting up automatic control of boiler. Energy, p. 280 (1970)
8. A.P. Kopelovich, Engineering methods of calculation of automatic regulators, *GNTI* (1960), p. 190
9. G.P. Pletnev, Computer-aided facilities management of TPP, Energoizdat, p. 361 (1981)
10. V.S. Mikhailenko, R.J. Harchenko, Using of hybrid networks in adaptive control systems of thermal power facilities. VNTU **1**, 1–9 (2012)

11. A. Jankowska, Neural models of air pollutants emission in power units combustion processes, in *Symposium on Methods of Artificial Intelligence* (2003), pp. 141–144
12. A.J. Leonenko, *Fuzzy Modeling in Matlab and fuzzyTech* (SPb.:BHV, St. Petersburg, 2003), p. 720
13. I.M. Sharovin, Industrial Controllers **2**, 27–32 (2010)
14. P. Yang, D.G. Peng, Y.H. Yang, Z.P. Wang, in *Proceedings of 2010 International Conference on Machine Learning and Cybernetics*, vol. 5 (2010), pp. 3300–3303
15. Monitoring and control of stoker-fired boiler plant using neural networks, UK Department of Trade and Industry, PS-156 (1999)
16. T. Takagi, M. Sugeno, IEEE Trans. Syst. Man Cybern. **15**, 116–132 (1985)

Profit Efficiency of Small Scale Layer Producers in Some Selected Local Government Areas in Sokoto State, Nigeria

L. Tanko, I.A. Nabil, and M.A. Maikasuwa

Abstract The study investigated the profit efficiency of small scale layer producers in some selected Local Government Areas (LGAs) in Sokoto State, Nigeria. The LGAs were purposively chosen based on high concentration of poultry farmers in these areas. A total of 120 respondents were selected randomly for this study. Data collection lasted from July to September, 2012. Data were analyzed using descriptive statistics, net farm income and stochastic frontier profit function models. Although layer enterprise was found to be a profitable venture, producers were found to be operating below the economic frontier given a mean profit efficiency value of 0.74 (i.e. 74 %) suggesting a scope for improvement by allocating existing resources more efficiently. Age, experience, credit, household size, gender and membership of association were found to be significant determinants of profit efficiency. The study recommends the stimulation of the domestic production of maize to curtail rising feed costs. There is also the need for extension education in the area of efficient resource management and cost saving strategies in layer production to raise farm incomes and profit.

Keywords Mean profit efficiency • Domestic production • Education • Cost saving strategies

L. Tanko (✉)
Department of Agricultural and Extension Technology, Federal University of Technology, P.M.B. 65, Minna, Niger State, Nigeria
e-mail: unekmelikita@futminna.edu.ng

I.A. Nabil • M.A. Maikasuwa
Usmanu Danfodiyo University, P.M.B. 2346, Sokoto, Nigeria

A.M. Gil-Lafuente and C. Zopounidis (eds.), *Decision Making and Knowledge Decision Support Systems*, Lecture Notes in Economics and Mathematical Systems 675, DOI 10.1007/978-3-319-03907-7_17, © Springer International Publishing Switzerland 2015

1 Introduction

The production of food in Nigeria has not increased at the rate that can meet the increasing population. While food production increased at the rate of more than 2.5 %, food demand increased at a rate of more than 3.5 % due to high rate of population growth of 2.83 % [1]. The apparent disparity between the rate of food production and demand for food in Nigeria has led to increasing food importation and high rates of increase in food prices. According to Wethli [2], poultry production is one of the most profitable agricultural enterprises and the accruing returns from the enterprise can be used to improve the life of rural dwellers. Agromisa [3] reported that the level of consumption of meat and other animal protein in Nigeria is estimated at about 8 g per caput per day, which is about 27 g less than the 35 g per caput minimum requirements recommended by the Food Agriculture Organization [4].

Olerede [5] observed that birds constitute over 90 % of the current national livestock population and are of appreciable economic and social value to the investors and consumers. Poultry products which are sold contribute about 15 % to the annual financial income of the household [6]. Poultry provide meat, egg, feather, manure (convertible to fertilizer and natural gas) to play an important role in the rural economy [7]. He affirmed that small scale layer production is an important element in diversifying income generation of the producers and increasing household food security. Evboumwan [8] reported that the commercialization of poultry keeping is a recent development in a humid tropical country like Nigeria.

The measurement of efficiency remains an important area of research both in developing and developed countries. Profit efficiency depends on market forces, which in turn are influenced by the sectoral and marketing policies of the country. Battese and Coelli [9] measured profit efficiency in which certain restrictions were imposed. Efficiency could be measured from a production function or a profit function approach. Determining the efficiency status of farmers is very important for policy purposes in an economy where technologies are lacking [10]. Adegeye [11] emphasized that subsistence oriented production especially among small scale farmers, poorly developed inputs and product markets, policy reversals, low investment in livestock enterprises, weakened extension services, poor utilization of superior varieties of poultry birds are some of the constraints to efficiency. The neglect of agriculture in Nigeria can be classified into severe, mild, chronic and transient. Nigerian agriculture continue to be neglected because of persistent dumping of cheap subsidized food imports from developed agriculture, weak agricultural stakeholders capacity prolonged political instability as a result of the discovery of petroleum and gas. The consequences of this neglect include food insecurity, food import tendency, rural unemployment, endemic poverty and stunted agro-industrialization.

The relationships between efficiency, market indicators and household characteristics have not been well studied in Nigeria. An understanding of these relationships could provide policy makers with information to design programmes that can contribute to measures needed to expand the food production potential of the country and better measures that can enhance agricultural efficiency can be implemented. This study was designed to ascertain the profitability and economic performance of small scale layer production in the study area and identify the determinants of profit efficiency of layer producers in the study area.

2 A First Approach

The popular approach to measure technical efficiency component is the use of frontier production function. However, it has been argued that a production function approach to measure efficiency may not be appropriate when farmers face different prices and have different factor endowments. This led to application of stochastic profit function models to estimate farm specific efficiency directly. Battese and Coelli [9] extended the stochastic production frontier model by suggesting that the inefficiency effects can be expressed as a linear function of explanatory variables, reflecting farm-specific characteristics. The advantage of this model is that it allows the estimation of farm specific scores and the factors explaining the efficiency differentials among farmers in a single stage estimation procedure. The stochastic profit function is defined as:

$$\pi_j = f\left(P_{ij}, Z_{kj}\right).Exp(V_i - U_i). \tag{1}$$

π_j-normalized profit of the jth farm and it is computed as gross revenue less valuable cost divided by the farm specific output price P, f represents an appropriate function (e.g. Cobb-Douglas, Trans-log etc), P_i is the price of the jth variable input faced by the jth farm divided by the price of unit of output, kj is level of kth fixed factors for the jth farm, V_i is a random variable which is assumed to be $N(O, \delta v^2)$ and independent of U_i which are non-negative random variables which are assumed to be $N(O, \delta v^2)$ i.e. half normal distribution or have exponential distribution. If $U_j = 0$, the firm is operating on the frontier, obtaining maximum profit given the prices it faces and levels of fixed factors. If $U_j > 0$, the firm is inefficient. Profit efficiency is defined as:

$$\text{Profit efficiency} = \pi/\pi * f\left(P_{ij}, Z_{kj}\right)exp(V_i - U_i)/f\left(P_{ij}, Z_{kj}\right)expV_j$$
$$= exp\left(-u_j\right). \tag{2}$$

π is the observed profit and π^* is the frontier profit defined in terms of ratio of the observed profit to the corresponding frontier profit given the prices and the levels of fixed factors of production of the farmer.

The Empirical Model: the Cobb-Douglas stochastic frontier profit functional form is specified as:

$$\ln \pi_j = \ln \alpha_0 + \sum_{i=1}^{4} a_i \ln P_{ij} + \sum_{k=1}^{2} a_k \ln z_{kj} + V_j - U_j. \tag{3}$$

Where i refer to variable inputs, k refers to fixed inputs and j refers to farms respectively. π_j is the normalized profit in naira per layer enterprises defined as gross revenue less total variable cost divided by the price of layer. V is the normal random errors which are assumed to be independent and identical distributed having zero mean and constant variance. U is the non-negative random variable associated with the profit efficiency of the enterprises.

Factors believed to affect the profit efficiency of the broiler farmer were incorporated into the model and estimated jointly. The efficiency component model is specified as:

$$-U = b_0 + b_1X_{1j} + b_2X_{2j} + b_3X_{3j} + b_4X_{4j} + b_5X_{5j} + b_6X_{6j} + b_7X_{7j} + b_8X_{8j}. \quad (4)$$

Where U is the profit efficiency, X_{1j}–X_{2j} are factors believed to affect the level of profit efficiency of the farmer and b_0, b_1–b_8 are maximum likelihood estimates to be to be measured.

3 Conceptual Framework

3.1 Study Area

The study was conducted in Sokoto state, Nigeria. The state was created in January 1976 with the headquarters in Sokoto. It is made up of 23 Local Government Areas (LGAS) covering a total land area of 26,648,480 square kilometers, Metrological Stations in Nigeria State [12]. It shares common borders with Niger republic to the north, Kebbi state to the west and Zamfara state to the east. The state is located within latitudes 11° 30′–13° 50′ N and longitudes 4° 07′–6° 56′ E. The state has a population of 4,244,399 people based on 2006 population census [13]. The estimated population by 2012 was 11,844,317 inhabitants. The rainfall starts late (May) and ends early (September/October) with mean annual falls ranging between 130 and 500 mm [12]. The mean minimum and maximum temperature are 23–43 °C [12].

Livestock farming and arable crop production are the major occupations of the people in the state. The people of the state are involved in the production, harvesting and marketing of farm products and the main livestock reared included: poultry, cattle, sheep and goats.

3.2 Sampling Technique and Sample Size

The study was conducted in three selected local government areas (LGAs) of Sokoto state, namely, Sokoto North, Sokoto South and Wamakko. The LGAs were purposively chosen based on the high concentration of the population of poultry farmers in these areas which also was related to its cosmopolitan nature and high human population which drives the demand for poultry products. A total of 40 respondents were randomly selected from each of these local government areas, to make up a sample size of 120 respondents. A sample frame which denoted the list of layer producers in the selected local government areas was obtained from the State Ministry of Agriculture. Agricultural Development Programme (ADP) agents, resident in each of these locations were adequately trained on the type of data required to be elicited from the respondents and co-opted in the data collection process for this study. They assisted the researcher in eliciting relevant information using the questionnaire as data collection instrument. Primary data were elicited from the respondents with the aid of a structured questionnaire. Secondary information were also obtained from journals, previous research works and textbooks.

The information elicited from the respondents include, socio-economic characteristics of respondents such as age, marital status, level of education, years of experience in the business, participation in cooperative society, extension visit, access to credit, source of funds, number of birds, feeds, access to credit, labour used and management system as well as information on other quantitative variables of interest such as, production inputs, outputs and their respective prices. The economic variables considered for estimating efficiency of layer production are: price of day old chicks (in Naira), price of feeds (Naira), price of drugs/medication (Naira), price of family labour (Naira), transportation cost (Naira), price of hired labour (Naira), Annual depreciation on durable capital items (Naira), and number of birds raised (Number), etc. Data collection lasted from July to September, 2012.

3.3 Methods of Data Analysis

The analytical techniques used in this study include, descriptive statistics such as means, frequency distributions and percentages to achieve objectives 1 and 4 respectively. The second objective was achieved by using the farm budget model. Objective three was achieved using the Cobb Douglas transcendental stochastic frontier profit function model. The farm budget model is a tool used to determine the level of resources used and output realized in farm enterprises [14]. The farm budget model was used to ascertain the profitability of small scale layer producers in the study area. The farm budget model is defined as:

$$NFI = GI - TC. \tag{5}$$

Where NFI = Income or profit (refers to the difference between gross income and total of costs of layer production in the study area), GI = Gross income represents the sum of total value of layer production (sales of egg, birds and poultry droppings), TC = Total cost refers to all the expenses incurred in the layer production by the farmer. These include fixed costs and variable costs, TVC = Total variable costs. These are costs that vary according to expenditure, incurred on variable inputs employed in production and TFC = Total fixed costs. These are the depreciation cost incurred on fixed inputs used during the production period.

$$NFI = GI - (TVC + TFC). \tag{6}$$

$$NFI = GI - TVC - TFC. \tag{7}$$

$$NFI = \sum_{j=1}^{m} P_j Q_j - \sum_{k=1} P_k Q_k - TFC. \tag{8}$$

Where, Pj = price of unit of Jth output, Qj = quantity of Jth output, Pk = price of Kth input, Qk = quantity of Kth input, $\sum$ = summation sign, TVC = total variable cost, TFC = total fixed cost, other variables are as previously defined.

3.4 Specification of the Stochastic Frontier Profit Function

The Cobb-Douglas transcendental logarithmic profit function was used to identify the determinants of profit efficiency of small-scale layer producers in the study area. The model estimated by jointly combining the production as well as inefficiency factors in a single stage maximum likelihood estimation procedure using computer software frontier version 4.1 [9] to identify the determinants of profit efficiency. The model is explicitly specified as:

$$
\begin{aligned}
\ln \pi = {} & \ln b_0 + b_1 \ln X_1 + b_2 \ln X_2 + b_3 \ln X_3 + b_4 \ln X_4 + b_5 \ln X_5 + \\
& b_6 \ln X_6 + b_7 \ln X_7 + b_8 \ln X_8 + 0.5b_{11} \ln X_1^2 + 0.5b_{22} \ln X_2^2 + \\
& 0.5b_{33} \ln X_3^2 + 0.5b_{44} \ln X_4^2 + 0.5b_{55} \ln X_5^2 + 0.5b_{66} \ln X_6^2 + \\
& 0.5b_{77} \ln X_7^2 + 0.5b_{88} \ln X_8^2 + b_{12} \ln X_1 \ln X_2 + b_{13} \ln X_1 \ln X_3 + \\
& b_{14} \ln X_1 \ln X_4 + b_{15} \ln X_1 \ln X_5 + b_{16} \ln X_1 \ln X_6 + b_{17} \ln X_1 \ln X_7 + \\
& b_{18} \ln X_1 \ln X_8 + b_{23} \ln X_2 \ln X_3 + b_{24} \ln X_2 \ln X_4 + b_{25} \ln X_2 \ln X_5 + \\
& b_{26} \ln X_2 \ln X_6 + b_{27} \ln X_2 \ln X_7 + b_{28} \ln X_2 \ln X_8 + b_{34} \ln X_3 \ln X_4 + \\
& b_{35} \ln X_3 \ln X_5 + b_{36} \ln X_3 \ln X_6 + b_{37} \ln X_3 \ln X_7 + b_{38} \ln X_3 \ln X_8 + \\
& b_{45} \ln X_4 \ln X_5 + b_{46} \ln X_4 \ln X_6 + b_{47} \ln X_4 \ln X_7 + b_{48} \ln X_4 \ln X_8 + \\
& b_{56} \ln X_5 \ln X_6 + b_{57} \ln X_5 \ln X_7 + b_{58} \ln X_5 \ln X_8 + b_{67} \ln X_6 \ln X_7 + \\
& b_{68} \ln X_6 \ln X_8 + b_{78} \ln X_7 \ln X_8 + V - U
\end{aligned} \tag{9}
$$

Where π = Net profit, ln = Natural logarithm, X_1 = Price of day old chicks (Naira), X_2 = Price of Feeds/Feeds Supplements (Naira), X_3 = Price of Drugs/Veterinary services (Naira), X_4 = Price of family labour (Naira), X_5 = Price of Transportation (Naira), X_6 = Price of hired labour (Naira), X_7 = Annual depreciation on durable capital items (Naira), X_8 = Number of birds (Number), b_0 = Constant, b_1–b_{78} = Maximum likelihood estimates, V = Statistical disturbance term and U = Farmer specific characteristics related to production efficiency, where:

$$U = \beta_0 + \beta_1 Z_1 + \beta_2 Z_2 + \beta_3 Z_3 + \beta_4 Z_4 + \beta_5 Z_5 + \beta_6 Z_6 + \beta_7 Z_7 + \beta_8 Z_8 + \beta_9 Z_9. \tag{10}$$

And β_0 = Constant term, $\beta_1 - \beta_9$ = Maximum likelihood estimates to be measured, Z_1 = Age of farmer (Years), Z_2 = Educational level (Number of years spent in school), Z_3 = Farming experience (Years), Z_4 = Farm size (Measured by total number of birds), Z_5 = Access to credit (Naira), Z_6 = Number of extension visits, Z_7 = Farm household size (Number of family members), Z_8 = Gender (male = 1, female = 2), Z_9 = Membership of association (member = 1, non-member = 0).

4 Results and Discussion

4.1 Socio-economic Characteristics of Respondents

The distribution of the respondents according to socioeconomic characteristics is presented in Fig. 1. The results of the study presented in Fig. 1 revealed that majority of the respondents (70 %) were males, while the remaining 30 % were females. This indicates that layer production in the study area was dominated by male. The implication of male dominance may be that productivity will be higher, because, males have the tendency to be more labour efficient as compared to their female counterparts. However, this finding underscores the need for the design of policy to take into account this gender related peculiarity. Taking labour efficiencies into concern, this findings supported Reddy [15] who found that the male folk were more profit efficient as compared to their female counterparts in the study area.

Results in Fig. 1 further revealed that majority of the respondents were married and a typical respondent was married and 33 years of age with five family members. Age affects decision and actions made in agriculture, because people's thoughts, behavior and needs are primarily related to their ages [16]. The results show that majority of the farmers are relatively young and are still in their agriculturally active age bracket. The implication is that younger farmers are likely to adopt innovation faster than older ones. The finding is in agreement with Sani [17]. All the respondents had one form of education or the other. Studies have revealed that education influences the adoption of improved practices of modern agriculture [18]. An educated person is more likely to adopt modern farming practices more easily and hence could be a better producer.

The findings of this study also indicated that a typical poultry farmer had 5 years of experience in the business. This finding also agrees with Oluwatayo [19] who found that farmers in Ekiti State also had up to 7 years of experience in the business. Farmers with more years of experience in an enterprise would be more likely to be efficient, may possess realistic planning imperatives and may have better knowledge of climatic conditions and market situations.

Fig. 1 Distribution of the respondents according to socioeconomic characteristics. Source: Field survey, 2012

Sex	Frequency	Percentage
Male	84	70.00
Female	36	30.00
Total	120	100.00
Marital Status		
Married	97	80.80
Single	23	19.20
Total	120	100.00
Education level		
Quranic	13	10.80
Primary	2	1.70
Secondary	12	10.00
Tertiary	93	77.00
Total	120	100.00
Extension access		
Yes	67	55.80
No	53	44.20
Total	120	100.00
Credit access		
Yes	63	52.50
No	57	47.50
Total	120	100.00
Source		
Self financing	48	40.10
Cooperative society	69	57.50
Commercial bank	1	0.80
Friends/relatives	1	0.80
Money lenders	1	0.80
Total	120	100.00
Management system		
Intensive	114	95.00
Semi-intensive	2	1.70
Extensive	4	1.30
Total	120	100.00
Labour		
Family	52	43.30
Hired	68	56.70
Total	120	100.00

	Minimum	Maximum	Mean	Std. error	Std. dev.
Age	20.00	60.00	33.00	0.73	8.00
Experience	1.00	21.00	5.00	0.31	3.42
Household size	1.00	15.00	5.00	0.34	3.75

4.2 Profitability of Layer Production Enterprises

The costs and returns of layer production in Naira per bird are presented in Table 1. Results in Table 1 indicated that an entrepreneur spent a total of ₦845.73 as total variable costs and ₦49.57 as total fixed costs to produce a bird and realized a total gross income of ₦2,688.71 for every bird raised up to laying in the study area. The net profit realized per bird in the study area was ₦1,793.41/bird. The variable cost items indicated that feed accounted for the highest contribution to the total cost of production which was 43.95 % of the total costs. The results also revealed that acquisition of foundation stock ranked second highest representing 26.74 % of the total costs of producing a layer bird. Vaccines and medication ranked third in decreasing magnitude of importance with 8.06 %. The results further indicated that for the revenue items, proceeds from sale of egg generated more revenue as compared to sales from spent layer and droppings respectively. Proceeds from the sale of eggs accounted for 59.12 % of the total gross income realized per layer, while sale of spent layer hen accounted for 25.52 % and droppings 15.36 % of the total gross income per layer produced. This implies that feed is an essential cost item in layer production. This agrees with Intisar [20], Sharabeen [21], Yusuf and Malamo [22] and Adepoju [23] who also found that feed cost comprised the highest share in the total cost of poultry production.

The results also showed that majority of the revenue is generated from the sale of eggs. This is similar to the findings of Narahari [24], Rajendran and Samarendu [25] and Emam and Hassan [26] who found that sale of eggs contributed the highest share of the total revenue realized by egg producers. The result further showed that the average gross income per bird was N2,688.71 and net income was N1,793.41. This implies that layer production was profitable in the study area. The results also agree with the finding of Reddi [15] and Rajendran and Samarendu [25] who found that gross margins and net returns increases with increase in farm size and was profitable. It also lent credence to the findings of Yusuf and Malomo [22], Sani [17] and Rajendran and Samarendu [25] who opined that many factors affect the profitability which may include cost of birds, price of egg among others depending on the location of the farm.

4.3 Financial Analysis

Financial analysis was done to assess the economic performance of layer enterprises in the study area. Results in Table 1 show the computed farm financial ratios of layer producers in the study area. The investment turnover and simple rate of return ratios were 54.12 and 2.00 respectively. This indicates that for a typical respondent, the business made returns of 54.12 % or ₦54.12 kobo on every Naira invested on the farm. Olukosi and Erhabor [34] posited that the higher the rate of returns on capital, the better it is for the success of the farm business. A simple rate

Table 1 Costs and returns of layer production (in Naira per bird)

Cost items	Amount (N)/bird	Percentage of total
Variable costs		
(i) Foundation stock	239.36	26.74
(ii) Feeds and supplements	393.49	43.95
(iii) Vaccines and medication	72.19	8.06
(iv) Electricity	27.91	3.12
(v) Water	28.47	3.18
(vi) Salaries	35.06	3.92
(vii) Repairs and maintenance	18.69	2.09
(viii) Transport	13.28	1.48
(ix) Litter material	9.45	1.06
(x) Commission/tax	7.84	0.88
Total variable costs(TVC)	845.73	94.46
Fixed cost (Depreciation on fixed assets such as feeders, drinkers, interest payments, etc)	49.57	5.54
(C) Total cost (TC)	895.30	100.000
(D) Revenue		
(i) Egg	1,589.54	59.12
(ii) Culled layer	686.16	25.52
(iii) Droppings	413.00	15.36
Total gross income (TGI)	2,688.71	100.00
(E) Net Profit(TGI − TC)	1,793.41	
(F) Farm financial ratios		
(i) Investment turn over	54.24	
(ii) Simple rate of return	2.00	
(iii) Operating ratio	0.31	
(iv) Income/expenses ratio	3.00	

Source: Field survey, 2012

of returns ratio greater than 1 is acceptable for a farm business. The operating ratio was 0.31. It shows the proportion of the gross income used to off-set the operating costs. The operating cost is directly related to the farm variable input usage. Generally, the lower the value of the operating ratio, the better is the financial position of the farm. An operating ratio of 1 means that the gross income barely covers the expenses on the variable inputs used on the farm or the enterprise is at the break-even point. Results further show an income/expenses ratio of the enterprise of 3.00. The ratio was used to measure the value of the output exceeding the total cost of production. The ratio of 3.00 so obtained shows that the revenue exceeded the total costs of production. This indicates that the business is solvent. Income-expenses ratio of large magnitude suggests that the enterprise is in a better the financial position.

4.4 Estimated Profit Function

The Maximum Likelihood Estimates of the stochastic frontier profit function for layer enterprise are presented in the Table 2. The result shows that the sigma squared (δ^2) was 0.56 and significant at 0.01 probability level. This implies the correctness of the specified distributional assumptions about the error term. The gamma (γ) value was 0.99 and significant at 0.01 probability level, suggesting that the sources of error were stochastic. Individual profit efficiency indices ranged between 0.10 (10 %) and 0.99 (99 %) with a mean value of 0.74 (74 %). This shows a wide gap between the most economically efficient farmer and the worst farmer in the sample given the average profit efficiency value of 0.74 (74 %). This implies that there is 26 % unattained efficiency for a typical layer producer. There is therefore need for improvement in efficiency by optimally allocating existing resources given input-output prices to move production to the frontier.

The elasticity estimates of price of day old chicks, feed supplements, drugs and veterinary services, price of transportation, price of hired labour and the number of birds raised were statistically significant at 1 % levels respectively. The price of day old chicks had a coefficient of 1.5472. This implies that an increase in the price of day old chicks by 1 %, holding other variables constant, would lead to an increase in the profit obtained from egg production by 1.5472 %. The estimated coefficients with respect to price of feeds/supplement, price of drugs/veterinary services, transportation , hired labour, annual depreciation and number of birds were −0.6380, −0.9058, 1.1601, 0.4286, −0.6990 and −0.0449 respectively. This also implies that a 1 % increase in each of these variable inputs had led to an increase (if the coefficient was positive) or a decrease (if the coefficient was negative) in the amount of profit realized respectively. The magnitudes of estimated coefficients are of economic relevance. The estimated coefficient of feed variable was −0.6380 and statistically significant at 0.01 probability level. Adepoju [23] stressed that feed cost is the most important single cost item associated with layer production due to increase in the cost of maize, groundnut cake, soya bean meal, fish meat and scarcity of wheat and corn offal. Sonaiya and Swan [27] earlier noted that availability of feeds at economic prices is by far the most important condition for profitable layer production, because, it constitutes more than 75 % of the total expenditure. Further lending credence to this assertion, Adeyinka and Mamman [28] observed that the cost of feed as a percentage of total variable cost was 67.8 % in layer enterprise and concluded that upward increase in feed cost led to reduced profit margin. The estimated coefficient for price of drugs and veterinary service was negative and significant at 0.01 probability level. They also found out that the cost of drugs/veterinary services accounted for 5–10 % of the total variable cost. Increase in the cost of drugs makes it difficult to check the increase in mortality rate as it depresses farm profits.

The parameter estimate of hired labour was positive and significant at the 0.01 probability level. Dillion [29] reported that the cost of labour constituted the second largest after feed. It is therefore pertinent that the efficiency of labour should be as

Table 2 Maximum likelihood Estimates of production factors

Variables	Parameter	Coefficient	Standard error (SE)	t-ratio
Constant	b_0	2.45	1.04	2.36**
Price of day old chicks	b_1	1.55	0.26	5.88***
Price of feed/supplement	b_2	−0.64	0.21	−3.05***
Price of drugs/veterinary services	b_3	−0.91	0.27	−3.34***
Price of family labour	b_4	−0.12	0.09	−1.31
Price of transportation	b_5	1.16	0.24	4.87***
Price of hired labour	b_6	0.43	0.14	3.11***
Annual depreciation	b_7	−0.70	0.19	−3.37***
Number of birds	b_8	−0.04	0.08	−5.89***
Interaction terms				
Price of day old chicks × price of day old	b_{11}	−1.54	0.33	−4.65***
chicks	b_{22}	−0.40	0.12	−3.35***
Price of feed/supplement × Price of feed/	b_{33}	0.13	0.21	0.66
supplement	b_{44}	−0.09	0.03	−0.31
Price of drugs/vet.Service × Price of drugs/	b_{55}	−7.62	0.11	0.67
vet.Service	b_{66}	0.04	0.04	0.90
Price of family labour × Price of family	b_{77}	−0.80	0.12	−6.41***
labour	b_{88}	−0.26	0.27	−0.97
Price of transportation × Price of	b_{12}	−0.51	0.30	−1.77*
transportation	b_{13}	2.24	0.17	3.06***
Price of hired labour × Price of hired labour	b_{14}	−0.21	0.14	−1.44
Annual depreciation × Annual depreciation	b_{15}	0.19	0.27	0.71
Number of birds × Number of birds	b_{16}	−0.14	0.24	−0.59
Price of day old chicks × Price of feed/	b_{17}	0.54	0.38	1.42
supplement	b_{18}	0.61	0.34	1.82*
Price of day old chicks × Price of drugs/vet.	b_{23}	−0.20	0.18	−1.16
Services	b_{24}	0.32	0.14	2.46**
Price of day old chicks × Price of family	b_{25}	−7.00	0.27	−0.38
labour	b_{26}	0.28	0.24	1.60
Price of day old chicks × Price of	b_{27}	0.92	0.38	2.94***
transportation	b_{28}	0.76	0.34	1.58
Price of day old chicks × Price of hired	b_{34}	0.20	0.18	2.07**
labour	b_{35}	0.40	0.13	4.58***
Price of day old chicks × Annual	b_{36}	0.19	0.19	1.39
depreciation	b_{37}	−0.96	0.18	−7.78***
Price of day old chicks × Number of birds	b_{38}	−1.83	0.31	−3.76***
Price of feed/supplement × Price of	b_{45}	0.12	0.48	1.20
drugs/vet.Services	b_{46}	−0.33	0.10	−2.60**
Price of feed/supplement × Price of family	b_{47}	−0.49	0.09	−6.81***
labour	b_{48}	0.55	0.14	3.83***
Price of feed/supplement × Price of	b_{56}	7.08	0.12	0.68
transportation	b_{57}	−0.13	0.49	−0.45
Price of feed/supplement × Price of hired	b_{58}	−0.24	0.10	−0.58
labour	b_{67}	−0.55	0.13	−0.52
Price of feed/supplement × Annual	b_{77}	0.53	0.14	2.24**
depreciation	b_{78}	1.14	0.15	3.05***
Price of feed/supplement × Number of birds				
Price of drugs/vet. services × Price of family labour				

(continued)

Table 2 (continued)

Variables	Parameter	Coefficient	Standard error (SE)	t-ratio
Price of drugs/vet. services × Price of transportation				
Price of drugs/vet. services × Price of hired labour				
Price of drugs/vet. services × Annual depreciation				
Price of drugs/vet. services × Number of		0.56		6.04***
birds		0.99		5.94***
Price of family labour × Price of				6.04***
transportation				5.94***
Price of family labour × Price of hired		0.99		
labour		0.10		
Price of family labour × Annual		0.74		
depreciation				
Price of family labour × Number of birds				
Price of transportation × Price of hired labour				
Price of transportation × Annual depreciation				
Price of transportation × Number of birds				
Price of hired labour × Annual depreciation				
Price of hired labour × Number of birds				
Annual depreciation × Number of birds				
Diagnostic Statistics				
Sigma-squared (δ^2)				
Gamma (γ)				
Log-likelihood function				
LR-Test				
Maximum value				
Minimum value				
Mean profit efficiency				

Number of observation = 120

Source: computer printout of frontier version 4.1/field survey, 2012

*significant at 10 %, **significant at 5 %, ***significant at 1 %

high as possible if profit is to be maximized in the enterprise. The estimated coefficient of capital inputs was −0.9058 and statistically significant at 0.01 probability level. Obinne [18] stressed that fixed costs affect the profit of most crops and livestock enterprises especially in the short-run planning period. Expenditure on fixed cost items depends on the size of the farm and not on the output level in which the enterprise is operating. The estimated coefficient of day old chicks is 0.4286 and statistically significant at the 0.01 probability level. This implies that increasing the number of day old chicks will lead to increase in profit. The results agrees with the findings of Gueye [30] that day old chicks require less feeds, drugs, medication and labour.

4.5   Determinants of Profit Inefficiency

The determinants of profit efficiency in layer production enterprise in the study area are presented in Table 3. The results indicates that age of farmer (-0.198), farming experience (0.0515), access to credit (-0.0711), farm household size (0.0448), gender (0.0423) and membership of association (0.0819). were the factors that significantly influenced the profit efficiency of layer producers in the study area. The negative and significant coefficient of age of farmer, farming experience and access to credit indicates that an increase in these variables resulted in a reduction in the levels of profit efficiency or increase in profit inefficiency of the respondents. This finding is consistent with the findings of Rahman [31] and Oluwatayo [19]. However, with respect to years of formal education, this finding is not consistent with *a priori* expectation of the sign of the estimated coefficient. This is because, education is expected to exert a positive influence on the profit efficiency of the farmer. Farmers having more years of formal schooling are expected to be more profit efficient. Similarly, the negative sign of the coefficient for credit was also contrary to *a priori* expectation of a positive sign for this variable. This is because, access to credit provides the farmer with means of expanding and improving his farm. Hence, lack of access and utilization of credit facility will exert a negative effect on profit efficiency. Okoh [32] corroborates this fact by reporting that credit increases the net revenue obtained from fixed inputs, market conditions and individual characteristics, while credit constraints decreases the efficiency of farmer by limiting the adaptation of high yielding varieties and the acquisition of information needed to increase productivity. Another possible implication could be that farmers allocated part of the loan acquired to finance non-agricultural ventures which depressed their efficiency.

The result of participation in cooperative society with a coefficient of 0.0819 implies that membership of organized farmer groups boosted the level of profit efficiency of the respondents. A poultry egg farmer who participated more in farmers cooperative would be more likely to be profit efficient as compared to his/her counterpart who did not. Through non-formal interpersonal contacts, farmers tend to learn better production techniques as compared to when they operate in isolation. They are more likely to imbibe the success story and best practices of their counterparts to enhance efficiency. This finding agrees with Okoh [32] and also supports the findings of Wainaina [33] who found that participation in cooperative activities may give room to assessing credit facility to enhance their productivity. The group is likely to hedge the collateral requirement for the prospective borrower to finance production activities.

The result of age of farmers implies that age of farmers influences profit efficiency adversely. In other words, as the age of the farmer increases, his/her level of profit efficiency declines. This result does not agree with the findings of Simsek and Karkacur [16] that age enhanced productivity as decisions and actions made is related to their ages. The result of household size suggests that large household size comprising of persons of active age bracket could constitute a

Table 3 Determinants of inefficiency in layer production

Variables	Parameter	Coefficient	Standard error (SE)	t-ratio
Constant	β_0	1.42	0.95	1.50
Age of farmer	β_1	−0.09	0.02	−4.13
Educational level	β_2	−0.01	0.02	−0.71
Farming experience	β_3	0.05	0.02	2.79
Farm size	β_4	0.01	0.01	1.49
Access to credit	β_5	0.07	0.02	4.25
Number of extension visits	β_6	−0.02	0.01	−1.67
Farm household size	β_7	0.04	0.01	4.21
Gender	β_8	0.04	0.01	3.98
Membership of association	β_9	0.08	0.04	2.27

Source: Field survey data 2012

work force with the propensity to increase efficiency [19]. The result of the gender coefficient was positive. This suggests that male farmers are economically more efficient than female farmers in the study area and the enterprise is therefore stereotyped to sex. This agrees with the findings of Oluwatayo [19] who found that male farmers were more efficient than their female counterparts. Due to this fact, they were favoured in terms of access to extension services, credit access and training schemes, farm input supplies and services and new production technologies.

5 Conclusion and Policy Recommendations

The results of this study indicated that layer production is a profitable venture but producers are operating below the economic frontier even though the financial analysis indicated solvency. Marked mal-allocation of existing resources was responsible for sub-optimal efficiency by entrepreneurs. To provide a framework for the way forward, the following recommendations are suggested and the relevant actors identified for implementing these recommendations. There is need for cost-saving by the entrepreneurs to enable them operate on the frontier. It is recommended that lending rates to the agricultural sector should be below two digits in order to spur more and faster transformation of the poultry industry thereby contributing to realizing national development goals to overcome food insecurity, improve nutrition and overcome poverty. There is need to stimulate domestic production of maize and other feed crops in a bid to curtail rising feed costs. The government can do this by providing incentives and a ready market for producers of corn for domestic and industrial purposes. Such incentives should include single digit lending rates and the provision of fertilizer, a critical input in maize production. This would require that farmers belong to organized groups to enable them access production inputs. The existing farmer cooperative societies should be further strengthened to necessitate improved access to credit and other facilities.

There is also the need for extension education in the area of resource management and cost saving methodologies, feed formulation, purchasing quality breeds so as to improve farm income and profit. Poultry farmers should be encouraged to keep records for accountability and prudence. Veterinary services were found to be significant determinants of farm profit. A public information campaign aimed at ensuring public knowledge about poultry disease would greatly help to improve simple daily practices and reduce the risk of disease. Television would be the best way to inform people but the large majority of Nigerians particularly in rural areas have access only to radio. This should also be included in the extension education content for dissemination to poultry farmers.

References

1. CBN, *Economic and Financial Review* (Minting Press, Lagos, Nigeria, 2004), p. 80
2. E. Wethli, *Chickens for Profit Starting a Small Poultry Business* (ITDG, UK, 2005), p. 100
3. A. Agromisa, *Agrodok Small-Scale Chicken Production* (CTA, Wageningen, The Netherland, 2006), p. 91
4. Food and Agricultural Organization, Egg marketing. A guide for the production and sale of eggs. FAO Agricultural Services Bulletin 150, p. 10 (2003)
5. B. Olerede, Non-conventional feed stuffs in poultry Nutrition (2005)
6. J. Olukosi, O. Abraham, *Agricultural Production Economics, Principles and Application*, 3rd edn. (G. U. Publications, Abuja, Nigeria, 2008), p. 112
7. A. Monsi, in *Animal Science Workshop in the New Millennium: Challenges and Options*, 6 Nov 2005, p. 16
8. G. Evbuomwan, Economic and Financial Review, Central Bank of Nigeria (2006)
9. G. Battese, T. Coelli, Empir. Econ. **20**, 325–332 (1995)
10. D. Jirong, C. Lovell, P. Schmidt, J. Econ. **6**, 21–37 (1996)
11. V. Adegeye, Unpublished Ph.D. thesis, Department of Agric. Economics, University of Ibadan, Ibadan, 2003
12. Metrological stations in Nigeria states, The metrological stations in Nigeria states. The comparison of 2008/2009 means monthly maximum temperature and rainfall in Nigeria (2009), pp. 1–4
13. National Population Commission, *Provisional Census Figure for Sokoto State* (Federal Government of Nigeria, Abuja, 2006)
14. J. Olukosi, S. Isitor, *Introduction to Agricultural Marketing and Prices, Principles and Application*, 2nd edn. (G. U Publications, Abuja, Nigeria, 2005), pp. 89–91
15. R. Reddi, Unpublished Ph.D. thesis, Department of Poultry Science, Madras Veterinary College, Chennai, cited in K. Rajendran, M. Samarendu, Int. J. Poultry Sci. **2**(2), 153–158 (2003)
16. D. Simsek, O. Karkacur, in *The 15th International Congress of Mediterranean Federation for Health and Production of Ruminants* (1996)
17. R. Sani, S. Musa, M. Daneji, M. Yakasai, O. Ayodele, Continent. J. Agr. Econ. **1**, 14–19 (2007)
18. C. Obinne, Afr. J. Biotechnol. **7**(9), 1227–1286 (1991)
19. I. Oluwatayo, A. Sekumade, S. Adesoji, World J. Agr. Sci. **4**(1), 91–99 (2008)
20. Y. Intisar, Afr. J. Biotechnol. **5**(18), 2491–2496 (1995)
21. I. Sharabeen, cited in A. Emam, A. Hassan, Afr. J. Biotechnol. **5**(18), 2491–2496 (2010)
22. S. Yusuf, O. Malamo, Int. J. Poultry Sci. **6**(9), 627–629 (2007)
23. A. Adepoju, Int. J. Agr. Econ. Rural Dev. **2**(1), 7–14 (2008)
24. D. Narahari, Int. J. Poultry Sci. **2**(2), 153–158 (2003)

25. K. Rajendram, M. Samarendu, Int. J. Poultry Sci. **2**(2), 153–158 (2003)
26. A. Emman, A. Hassan, Afr. J. Biotechnol. **5**(18), 2491–2496 (2010)
27. E. Sonaiya, S. Swan, Animal Production and Health 1 (2004)
28. I. Adeyinka, L. Mamman, NAPRI, Ahmadu Bello University, Zaria (2002), pp. 68–75
29. J. Dillion, S. Bruno, J. Hardakeu, *Farm Management Research for Small Farmer Development* (Food and Agriculture organization of the United Nations, Rome, 1998)
30. E. Gueye, World Poultry Sci. J. **55**, 187–198 (1999)
31. S. Rahman, Ph.D. thesis, Department of Agricultural Economics and Rural Sociology, Ahmadu Bello University, Zaria, 2001 cited in S.A. Rahman, J.F. Alamu, Niger. J. Sci. Res. **4**(1), 45–49 (2003)
32. S. Okoh, S. Rahman, H. Ibrahim, Livestock Research for Rural Development, vol. 22, 160 (2010). http://www.irrd.org/irrd22/9/okoh22160.htm. Retrieved 30 Sept 2012
33. P. Wainaina, W. Okello, J. Nzuma, in *International Association of Agricultural Economists (IAAE) Triennial Conference*, 18–24 Aug 2012
34. J.O. Olukosi, P.O. Erhabor, *Introduction to farm management economics: principles and applications* (Agitab, Zaria, 1988)

Investigation into the Effect of Velocity Distribution on Dry Season Hydrocarbon Degradation in Pond System

Peter C. Ukpaka and Humphrey A. Ogoni

Abstract Investigation into the hydrocarbon degradation was examined by considering the effect of velocity distribution on substrate concentration and biomass built up as well as the physiochemical parameter of the pond system. The research work contains experimental and theoretical data. The model developed was simulated using the experimental data obtained from the field of investigation. The result obtained shown negative and positive velocity indicating that the negative velocity influences the inflow velocity distribution of the substrate as well as the physiochemical parameter concentration. The actual velocity operational on the pond system was evaluated by considering the different in positive (inflow) velocity from the negative (backflow) velocity. The research demonstrates the effect of velocity distribution on the substrate concentration, biomass build up, physiochemical parameters as well as evaluation of the theoretical maximum specific rated of microbial growth $\mu Tmax = 45.45 \times 10^{-4}$ day^{-1}, equilibrium constant for the microbial growth $KTm = 0.0125$, whereas the experimental values are given as $\mu Emax = 10.0 \times 10^{-5}$ (day)$^{-1}$ and $KEm = 200$. It is seen that velocity distribution in pond system influence the effectiveness of bio-remediation of hydrocarbon.

Keywords Hydrocarbon degradation • Physiochemical parameters • Bio-remediation of hydrocarbon

P.C. Ukpaka (✉)
Department of Chemical/Petrochemical Engineering, Rivers State University of Science &
Technology, Nkpolu P.M.B 5080, Port Harcourt, Rivers State, Nigeria
e-mail: chukwuemeka24@yahoo.com

H.A. Ogoni
Niger Delta University, Wilberforce Island, Bayelsa State, Nigeria

A.M. Gil-Lafuente and C. Zopounidis (eds.), *Decision Making and Knowledge Decision*   197
Support Systems, Lecture Notes in Economics and Mathematical Systems 675,
DOI 10.1007/978-3-319-03907-7_18, © Springer International Publishing Switzerland 2015

1 Introduction

The current worldwide complain against environmental degradation has attracted the attention of environmentalist, geoscientist, engineers etc. The exploration, production, transportation, refining and utilization, therefore have been a major player in environmental pollution. These activities have to be carried out in such a way as to protect and preserve the quality of the environment while still achieving desire economic benefits. In Nigeria, this is most pronounced in the Niger Delta area for both upstream and downstream sectors as their operations have rapidly increased. The essence of effluent water treatment is to protect the environment from further environmental degradation.

The present practice of most companies in wastewater disposal during exploration and production operation, to the receiving; lakes, rivers, ocean and sea without treatment are unacceptable. Consequently, the increase in petroleum and gas production and inadequate treatment implies that the future of man in these regions remains uncertain. Also, considering the current trends in Government environmental regulations regarding the disposal of wastewater (effluent water), there is urgent need to evaluate alternative and acceptable ways of reducing waste water disposal into the environment as well ascertains the effect of velocity distribution on the effectiveness of bioremediation in pond system. Bioremediation treatment techniques are in use to enhance environment clean up of polluted ponds, lakes, rivers, seas and the ocean. Field and laboratory investigations reveal that the treatment technique can be influenced by temperature, pH, dissolved oxygen, biological oxygen demand, chemical oxygen demand, moisture content, solubility of the contaminants etc. Grandy et al. [22]. Studies conducted in similar environments confirm the influence on the physicochemical properties on the biodegradation process [1, 2]. Field and laboratory studies were conducted to assess the performance of functional parameters in hydrocarbon degradation in pond system. The impact of continuous discharge of wastewater on the physicochemical properties of pond system was examined. The continuous discharge of wastewater in bioremediation technique in enhancing environmental clean-up. This arouses my interest to propose a multi-functional and multi-disciplinary approach in solving these environmental problems, such as possible factors that may influence bioremediation of polluted pond for continuous discharge of wastewater. This process may alter the hydrocarbon composition in the oxidation pond, and likely influencing the performance or activities of the microbes present in the oxidation pond. This is possible, because the change in the physicochemical properties of the hydrocarbon composition in the oxidation pond may result in increase or decrease in the microbial activities. Since the wastewater (produced water) is expected to be in excess of the volume required for pressure maintenance, the excess wastewater (produced water) must be disposed off through non-potable aquifers or by continued discharge into ponds, lakes, rivers, seas, or ocean etc. Although all the investigations conducted by various research groups revealed that intra-field disposal of excess wastewater (produced water) into the subsurface is the most permanent and acceptable method, the re-injection of wastewater may likely contaminate the underground water (portable water resources) [1]. This is the current approach employed by many of the oil and gas exploration and production companies located within the Niger Delta area of Nigeria [3].

The role of biodegradation in the chemical evolution of the residual petroleum hydrocarbon mixture has given rise to a new trend of technology in the petroleum industry [1]. The fundamental principles of this fast growing technology are to create conditions under which micro-organisms grow and use the petroleum hydro-carbon as substrate. The result of this is the transformation of the residual petroleum hydrocarbons discharged into a pond system to carbon dioxide and water. Microbial activities in petroleum hydrocarbon have long been reported. These activities have been noticed at various stages of the petroleum industry. To enhance the degrada-tion ability of microbes, various physicochemical properties of the medium (pH, temperature, dissolved oxygen and total dissolved solids) need to be monitored and controlled. The biodegradation rate of petroleum hydrocarbon, have been studied in relation to dissolution and spreading rate [4]. However, extensive literature survey reveals that no comprehensive kinetic model on the effect of momentum transfer have been presented for the biodegradation rate on microbial growth and decay rates of petroleum hydrocarbon residue [1, 2, 5–16].

Ponds have been in use in different parts of the world. The mechanism of effluent treatment in this technique involves the use of micro-organism to degrade the petroleum hydrocarbons present in the pond. Although petroleum hydrocarbon components released by the operations of the petroleum industry may enter into the environment and follow the normal biogeochemical cycle (anthropogenic), a lot needs to be done to finally clean up these petroleum hydrocarbons. At a time when environment quality and food productions are of major concern to mankind, a better understanding of the behaviours of petroleum hydrocarbons is required. Pollutant in the environment have been studied generally, (water, air, soil and even the subsur-face system) and its effect seems to be very significant [3, 8, 10–18] (Development of models for the prediction of functional parameters for hydrocarbon degradation in pond system).

However, in practice most system are multi-components, as in the case of relevant properties of petroleum hydrocarbons (linear alkanes—C_5 to C_9, C_{10} to C_{22}, C_{23} to C_{44}; branched chain alkanes; alkanes; cyclic alkanes; monoaromatics, BTEX; and polyaromatics 6–2 rings, 3–4 rings and <4 rings) as reported [19]. The conclusion of this investigation lack the effect of momentum transfer or continuous discharge of effluent in such process and above all kinetic model will be required to simulate the following in terms of stability in the pond system upon the influence of velocity distribution, biomass built up and physiochemical parameters of the pond system. The petroleum hydrocarbon composition and physicochemical properties of the pond system in all the conditions prevalent in Nigeria will be studied in relation with the effect of velocity distribution experienced due to continuous discharge of wastewater. On the other hand, the importance of microbial and substrate kinetic as well as product kinetic for anaerobic and aerobic reactors for wastewater treatment is well known by many researchers [17, 20].

The main objective of this study is for the development of models for the prediction of hydrocarbon degradation upon the influence of velocity distribution in pond system for continuous discharge of wastewater, as well as kinetic model of microbial metabolic activities for single and multiple enzymes catalysed and uncatalysed reactions, under the influence of velocity distribution. The objective of this study is to carry out a detailed investigation on the influence of velocity distribution and mass transfer on the functional parameters that attribute to the biodegradation of petroleum hydrocarbons in a continuous discharge of wastewater in pond system.

2   Materials and Methods

2.1   The Model

The general equation obtained in Eq. (1) was derived from the force balance model on a fluid element in a pond using the mathematical application method and principles of gravity friction and hydrostatic force. Applying the law of conservation of momentum equation, thus we have:

$$\frac{\partial u}{\partial t} + V\frac{\partial u}{\partial x} = gS_0. \tag{1}$$

The effect of velocity on hydrocarbon degradation was developed using the momentum transfer process for pond system which can be describe as a simple as a simple batch phenomenon under conditions where organic sedimentation, sediment reactions and loss of organic volatiles component of petroleum hydrocarbon are negligible. Therefore Eq. (1) can be resolved by the application of separation of variables

$$\frac{\partial u}{\partial t} + V_I\frac{\partial u}{\partial x} = gS_0.$$

Equation (1) is expressed by considering the following boundary condition such as:

$$\text{at } x_1 = 0, t = 0, U = C_n(2). \tag{2}$$
$$\text{at } x_1 = L, t = t(3). \tag{3}$$

Equation (1) can be further be expressed by using the necessary mathematical approach, such as:

$$\text{Let } U = T_1 x_1.$$

Where T is a function of (t) and x is a function of x only. The use of Eq. (3) called the method of separation of variable may enable one to reduce a partial differential equation to several ordinary differential equations. To this end we note:

$$\frac{\partial u}{\partial t} = T_1^1 x_1. \tag{4}$$

And

$$\frac{\partial u}{\partial t} = T_1 x_1^1. \tag{5}$$

Substituting Eqs. (3), (4) and (5) into Eq. (1) yields

$$T_1^1 x_1 + V_1 \left(T x_1^1 \right) = g S_0. \tag{6}$$

Dividing through Eq. (6) by $T_1 x_1$ yields

$$\frac{T_1^1}{T_1} + V_1 \frac{x_1^1}{x_1} = \frac{g S_0}{T_1 x_1}. \tag{7}$$

Equation (7) can be expressed by considering both sides to be constant. In practice, it is convenient to write this real constant as either λ^2 or $-\lambda^2$. Therefore Eq. (7) can be written as:

$$\frac{T_1^1}{T_1} = -V_1 \frac{T_1 x_1^1}{x_1} - \frac{g S_0}{T_1 x_1}. \tag{8}$$

$$\frac{T_1^1}{T_1} = -V_1 \frac{x_1^1}{x_1} = \frac{g S_0}{T_1 x_1} = \lambda^2. \tag{9}$$

Equation (9) is true only if $\lambda^2 > 0$. Therefore based on the assumption $\lambda^2 > 0$, Eq. (9) is expressed as:

$$\frac{T_1^1}{T_1} = \lambda^2. \tag{10}$$

Rearranging Eq. (9) yields

$$T_1^1 - \lambda^2 T_1 = 0. \tag{11}$$

Therefore, the general solution to Eq. (11) becomes;

$$T_1 = c_1 e^{\lambda^2 t}. \tag{12}$$

Similarly,

$$-V_1 \frac{x_1^1}{x_1} = \lambda^2. \tag{13}$$

Rearranging Eq. (13) yields:

$$-V_1 x_1^1 + \lambda^2 x_1 = 0. \tag{14}$$

Therefore, the general solution to Eq. (13) becomes;

$$x_1 = C_2 e^{\frac{\lambda^2}{V_1} x_1}. \tag{15}$$

Also, from Eq. (9) we have

$$\frac{gS_0}{T_1 x_1} = \lambda^2. \tag{16}$$

Therefore, Eq. (16) can be written as

$$\lambda^2 = \frac{gS_0}{T_1 x_1}. \tag{17}$$

Since time and distance is a function of velocity, therefore $T_I x_1$ can be written as V_I. Thus Eq. (17) becomes

$$\lambda^2 = \frac{gS_0}{V_I}. \tag{18}$$

Substituting Eqs. (12) and (13) into Eq. (3) yields

$$U_1 = \left(C_1 e^{\lambda^2 t} \right) \left(C_2 e^{\frac{\lambda^2}{V_I} x} \right). \tag{19}$$

Since $\lambda^2 = \frac{gS_o}{V_I}$ and $U_1 = T_1 x_1$ therefore Eq. (19) yields

$$U_1 = \left(C_1 e^{\frac{gS_0}{V_I} t} \right) \left(C_2 e^{\frac{gS_0}{V_I} x} \right). \tag{20}$$

Substituting the boundary conditions into Eq. (20) yields: for step 1, at $x = 0$, $t = 0$, $U = C_n$

$$U_1 = \left(C_1 e^{\frac{gS_0}{V_I}(0)}\right)\left(C_2 e^{\frac{gS_0}{V_I}(0)}\right).$$ (21)

$$C_n = C_1 C_2.$$ (22)

Therefore

$$C_1 = \frac{C_n}{C_2}.$$ (23)

For step II:
Considering the boundary condition, at $x = L$ and $t = t$. Therefore substituting the boundary condition in step II into Eq. (20) yields

$$U_1 = \left(C_1 e^{\frac{gS_0}{V_I}t}\right)\left(C_2 e^{-\frac{gS_0}{V_I}L}\right).$$ (24)

$$\text{Since } V_I = \frac{L_I}{t}.$$ (25)

Rearranging Eq. (25) yields

$$t = \frac{L_I}{V_I} \text{ and } L_I = V_I t.$$ (26)

Substituting Eq. (26) into Eq. (24) yields

$$U_{L(1)} = \left(C_1 e^{\frac{gS_0}{V_I}\frac{L_I}{V_I}}\right)\left(C_2 e^{-\frac{gS_0}{V_I}L_1}\right).$$ (27)

$$U_{L(I)} = \left(C_1 e^{\frac{gS_0}{V_I^2}L_I}\right)\left(C_2 e^{-\frac{gS_0}{V_I}L_I}\right).$$ (28)

Substituting Eq. (3.169) into Eq. (3.175) yields

$$U_{L(1)} = \left(\frac{C_n}{C_2} e^{\frac{gS_0}{V_I^2}L_I}\right)\left(C_2 e^{-\frac{gS_0}{V_I}L_I}\right).$$ (29)

Simplifying Eq. (29) reduces to:

$$U_{L(1)} = \left(C_n e^{\frac{gS_0}{V_I^2}L_I}\right)\left(e^{-\frac{gS_0}{V_I}L_I}\right).$$ (30)

Equation (30) can be written as

$$\text{In}\, \frac{U_L}{C_n} = \frac{gS_0 L_I (1 - V_I)}{V_I^2}. \tag{31}$$

$$\text{In}\, \frac{U_{L(I)}}{C_n} = \frac{gS_0 L_I (1 - V_I)}{V_I^2}. \tag{32}$$

Since

$$\lambda = \frac{gS_0}{V_I}.$$

Therefore Eq. (32) become

$$\text{In}\, \frac{U_{L(1)}}{C_n} = \frac{\lambda^2 L_I (1 - V_I)}{V_I}. \tag{33}$$

Rearranging Eq. (33) yields

$$\lambda^2 = \frac{V_I}{L_I (1 - V_I)}\, \text{In}\, \frac{C_n}{U_{L_I}}. \tag{34}$$

$$\frac{gS_0}{V_I} = \frac{V_I}{L_I (1 - V_I)}\, \text{In}\, \frac{C_n}{U_{L_I}}. \tag{35}$$

Therefore

$$S_0 = \frac{V_I}{gL_I (1 - V_I)}\, \text{In}\, \frac{C_n}{U_{L_I}}. \tag{36}$$

From Eq. (36) yields

$$\frac{V_I^2}{1 - V_I} = gS_0 L_I\, \text{In}\, \frac{C_n}{U_{L_I}}. \tag{37}$$

The developed model in Eq. (37) can be applied in monitoring the rate of degradation of the individual hydrocarbon, estimating the degree of influence of velocity distribution due to continuous discharge of wastewater and the affect area, estimating the spreading rate and diffusion rate for each hydrocarbon component in the oxidation pond system. The use of this Eq. (37) is useful for the prediction as well as the correlation of velocity and substrate concentration as a function of distance and time. The velocity $V_I^2/(1 - V_I)$ for the oxidation pond system was determined at the point of intercept on the $V_I^2/(1 - V_I)$ coordinate and the slope of the graph was used in the determination of gS_0 in C_n/U_{LI} for dry season.

2.1.1 Kinetics Model for Multiple Substrate and Single Enzyme Without Activator

In this study mathematical equations were developed for the multiple substrate and single enzyme reaction with activator and without activator, under the influence of velocity distribution due to continuous discharge of wastewater as well as considering when the velocity distribution in the process is negligible.

The pond system contains different mixture of petroleum hydrocarbons. Assuming 'n' number of petroleum hydrocarbons as the only source of carbon, using this as substrate, the reaction steps in the scheme involving multiple intermediate can be presented as follows:

$$[E] + [S_1] \overset{K_S}{\rightleftharpoons} [ES_1] \overset{K_{P_1}}{\rightarrow} [E] + [P_1]. \tag{38}$$

$$[E] + [S_2] \overset{K_S}{\rightleftharpoons} [ES_2] \overset{K_{P_2}}{\rightarrow} [E] + [P_2]. \tag{39}$$

$$[E] + [S_n] \overset{K_S}{\rightleftharpoons} [ES_n] \overset{K_{P_n}}{\rightarrow} [E] + [P_n]. \tag{40}$$

From Eqs. (38) to (39) are the reaction sequence for multiple substrate and single enzyme without activator. But, considering the reaction mechanism taking place in the pond in the presence of activator (high concentration of nitrogen and phosphorus effluent composition discharged into the pond) the following reaction sequence can be obtained as;

$$\begin{bmatrix} [E] + [S_1] & \overset{K_S}{\rightleftharpoons} & [ES_1] & \overset{K_P}{\rightarrow} & [E] + [P_1] \\ + & & + & & \\ [A] & & [A] & & \\ \uparrow\downarrow & & K_A \uparrow\downarrow & & \\ [EA] + [S_1] & \overset{\alpha K_S}{\rightleftharpoons} & [EAS_1] & \overset{\beta K_P}{\rightarrow} & [EA] + [P_1] \end{bmatrix} \tag{41}$$

$$\begin{bmatrix} [E] + [S_2] & \overset{K_{S2}}{\rightleftharpoons} & [ES_2] & \overset{K_P}{\rightarrow} & [E] + [P_2] \\ + & & + & & \\ [A] & & [A] & & \\ \uparrow\downarrow & & K_A \uparrow\downarrow & & \\ [EA] + [S_2] & \overset{\alpha K_S}{\rightleftharpoons} & [EAS_2] & \overset{\beta K_P}{\rightarrow} & [EA] + [P_2] \end{bmatrix} \tag{42}$$

$$\begin{bmatrix} [E] + [S_n] & \overset{K_s}{\rightleftharpoons} & [ES_n] & \overset{K_P}{\rightarrow} & [E] + [P_n] \\ + & & + & & \\ [A] & & [A] & & \\ \uparrow\downarrow & & K_A \uparrow\downarrow & & \\ [EA] + [S_n] & \overset{\alpha K_S}{\rightleftharpoons} & [EAS_n] & \overset{\beta K_P}{\rightarrow} & [EA] + [P_n] \end{bmatrix} \tag{43}$$

The mathematical expression for a typical aerobic pond reactor (batch reactor), the material balance can be expressed as; (without activator)

$$-\frac{1}{X}\frac{dS}{dt} = \mu^S \frac{X}{Y}. \tag{44}$$

Similarly, the mathematical expression for Monod equation for this typical aerobic pond reactor is given as:

$$\frac{dS}{dt} = \left(\mu_{max}^S \frac{S}{K_m + S}\right)\frac{X}{Y}. \tag{45}$$

The mathematical expression in terms of microbial substrate relationship is given as

$$Y = \frac{X - X_0}{S_0 - S} = \frac{dx}{dS}. \tag{46}$$

Substituting Eq. (46) into Eq. (45) in terms of Y yields

$$\frac{dS}{dt} = \left(\mu_{max}^S \frac{S}{K_m + S}\right)\frac{X}{dx/dS}. \tag{47}$$

The mathematical expression obtained in Eq. (47) is only for single component of the system. Therefore defining Eq. (47) in terms of multiple component system yields

$$\frac{dS}{dt} = \mu_{max}^S \left(\begin{array}{c} \dfrac{S_1}{K_{m_1} + S_1} \bullet \dfrac{S_2}{K_{m_2} + S_2} \bullet \dfrac{S_3}{K_{m_3} + S_3} \\[2mm] \bullet \dfrac{S_4}{K_{m_4} + S_4} \bullet \dfrac{S_5}{K_{m_5} + S_5} \bullet \bullet \bullet \dfrac{S_n}{K_{m_n} + S_n} \end{array}\right) \dfrac{X}{dX/dS}. \tag{48}$$

2.1.2 Kinetics Model for Multiple Substrates and Multiple Enzymes

These kinetic model investigations were geared towards their application in design and operation of petroleum hydrocarbon waste treatment system. From the research carried no kinetic model has been formulated for microbial growth in the biodegradation process of crude oil on the influence of velocity distribution due to continuous discharge of wastewater in the pond system. Consequently, this work is aimed at developing kinetic model of microbial growth on the biodegradation of crude oil for single and multiple enzymes under the influence of velocity distribution in the bioreactor.

In a multiple enzyme reaction the specific rate of biodegradation reaction in a pond system is expressed as the sum for the individual enzymes. Hence, the following reaction is obtained such as

$$[E_1] + [S_1] \overset{K_{S1}}{\rightleftharpoons} [E_1 S_1] \overset{K_P}{\rightarrow} [E_1] + [P_1].$$

$$[E_2] + [S_2] \overset{K_{S2}}{\rightleftharpoons} [E_2 S_2] \overset{K_P}{\rightarrow} [E_2] + [P_2].$$

$$[E_3] + [S_3] \overset{K_{S3}}{\rightleftharpoons} [E_3 S_3] \overset{K_P}{\rightarrow} [E_3] + [P_3].$$

$$[E_n] + [S_n] \overset{K_{Sn}}{\rightleftharpoons} [E_n S_n] \overset{K_P}{\rightarrow} [E_n] + [P_n].$$

The mathematical expression for the kinetic model for multiple substrate and multiple enzymes without activator is based on the Eqs. (44) (45), (46) and (47), but mathematical expression is given as the sum of or the individual enzymes present in the system and it is given as;

Rearranging Eq. (48) yields

$$\frac{dX}{dt} = \mu_{max}^{S} \left(\begin{array}{c} \dfrac{S_1}{K_{m_1} + S_1} \bullet \dfrac{S_2}{K_{m_2} + S_2} \bullet \dfrac{S_3}{K_{m_3} + S_3} \\ \bullet \dfrac{S_4}{K_{m_4} + S_4} \bullet \dfrac{S_5}{K_{m_5} + S_5} - - - \bullet \dfrac{S_n}{K_{m_n} + S_n} \end{array} \right) X. \qquad (49)$$

Similarly, Eq. (49) can further be rearranged to yield

$$\frac{dX}{X} = \mu_{max}^{S} \left(\begin{array}{c} \dfrac{S_1}{K_{m_1} + S_1} \bullet \dfrac{S_2}{K_{m_2} + S_2} \bullet \dfrac{S_3}{K_{m_3} + S_3} \\ \bullet \dfrac{S_4}{K_{m_4} + S_4} \bullet \dfrac{S_5}{K_{m_5} + S_5} - - - \bullet \dfrac{S_n}{K_{m_n} + S_n} \end{array} \right) dt. \qquad (50)$$

Integrating Eq. (50) yields

$$In \frac{X}{X_o} = \mu_{max}^{S} \left(\begin{array}{c} \dfrac{S_1}{K_{m_1} + S_1} \bullet \dfrac{S_2}{K_{m_2} + S_2} \bullet \dfrac{S_3}{K_{m_3} + S_3} \\ \bullet \dfrac{S_4}{K_{m_4} + S_4} \bullet \dfrac{S_5}{K_{m_5} + S_5} \bullet\bullet\bullet \dfrac{S_n}{K_{m_n} + S_n} \end{array} \right) t. \qquad (51)$$

$$X = X_0 \ell^{\mu_{max}^{S} \left(\begin{array}{c} \frac{S_1}{K_{m_1} +S_1} \bullet \frac{S_2}{K_{m_2} +S_2} \bullet \frac{S_3}{K_{m_3} +S_3} \\ \bullet \frac{S_4}{K_{m_4} +S_4} \bullet \frac{S_5}{K_{m_5} +S_5} \bullet\bullet\bullet \frac{S_n}{K_{m_n} +S_n} \end{array} \right) t}. \qquad (52)$$

Further simplification of Eq. (52) yields

$$\mu_{\max}^S t \left(\frac{S_1 S_2 S_3 S_4 S_5 - - - S_n}{K_{m_1} + S_1 \bullet K_{m_2} + S_2 \bullet K_{m_3} + S_3 \bullet K_{m_4} + S_4 \bullet K_{m_5} + S_5 \bullet \bullet \bullet K_{m_n} + S_n} \right)$$
$$= In \frac{X}{X_0}.$$

(53)

$$\mu_{\max}^S t = \frac{In \frac{X}{X_0} (K_{m_1} + S_1 \bullet K_{m_2} + S_2 \bullet K_{m_3} + S_3 \bullet K_{m_4} + S_4 \bullet K_{m_5} + S_5 \bullet K_{m_n} + S_n)}{S_1 S_2 S_3 S_4 S_5 \bullet \bullet \bullet S_n}$$
$$+ C.$$

(54)

If $X_o = 0$, and expressing Eq. (50) in terms of component i, yields

$$\mu_{\max}^S t = \left[\prod_{i=1}^{n} \left(\frac{(K_m)_i + [S]_i}{S_i} \right) \right] \frac{dX}{X}.$$

(55)

Integrating Eq. (55) yields

$$\mu_{\max}^S t = \left\{ \prod_{i=1}^{n} \left(\frac{(K_m)_i + [S]_i}{S_i} \right) \right\} In X + C_1.$$

(56)

Developing a mathematical model by considering the reaction mechanism involved from Eqs. (41) to (43), for typical aerobic batch reactor, the material balance could be expressed as:

$$\frac{dS}{dt} = \mu_A^S \frac{X}{Y}.$$

(57)

The mathematical expression for the Monod equation is given as

$$\frac{dS}{dt} = \left(\mu_A^S \frac{S}{K_m + S} \right) \frac{X}{Y}.$$

(58)

The mathematical expression for microbial-substrate relationship can be expressed as;

$$Y = \frac{X - X_o}{S_o - S} = \frac{dX}{dS}. \tag{59}$$

Substituting Eq. (58) into Eq. (59) yields

$$\frac{dS}{dt} = \mu^S_{A.max} \frac{S}{K_m + S} \bullet \frac{X}{dX/dS}. \tag{60}$$

For a multiple component systems

$$\frac{dX}{dt} = \mu^S_{A.max} \left(\frac{S_1}{K_{m_1} + S_1} \bullet \frac{S_2}{K_{m_2} + S_2} \bullet \frac{S_3}{K_{m_3} + S_3} \bullet \bullet \bullet \frac{S_n}{K_{m_n} + S_n} \right) \frac{X}{dx/dS}. \tag{60a}$$

Since the enzyme substrate complex [ES] and enzyme-activator-substrate complex [EAS] are both-product forming complexes and rate determining,

$$= K_P[ES] + \beta K_P[EAS]. \tag{61}$$

Similarly, a material balance for the enzyme can be expressed as;

$$[E_t] = [E] + [ES] + [EAS]. \tag{62}$$

Expressing each term in term of free enzymes

$$K_S = \frac{K_{-1}}{K_1} = \frac{[S][E]}{[ES]}. \tag{63}$$

$$[ES] = \frac{[S]}{K_S}[E]. \tag{64}$$

Similarly

$$\alpha K_A = \frac{[A][S]}{[EA]}[E]. \tag{65}$$

$$[EAS] = \frac{[A][S]}{\alpha K_A}[E]. \tag{66}$$

Substituting Eq. (60a) into (66) and rearranging yields

$$\frac{dS}{dt} = \mu_{A.\max}^{S} \left(\frac{\dfrac{[S_1]}{K_{m_1}\left(\dfrac{1+\dfrac{[A]}{K_A}}{1+\beta[A]/_{\alpha K_A}}\right) + [S_1]\left(\dfrac{1+\dfrac{[A]}{\alpha K_A}}{1+\beta[A]/_{\alpha K_A}}\right)} \cdot}{\dfrac{[S_2]}{K_{m_1}\left(\dfrac{1+\dfrac{[A]}{K_A}}{1+\beta[A]/_{\alpha K_A}}\right) + [S_2]\left(\dfrac{1+\dfrac{[A]}{\alpha K_A}}{1+\beta[A]/_{\alpha K_A}}\right)} \cdot}{\dfrac{[S_2]}{K_{m_n}\left(\dfrac{1+\dfrac{[A]}{K_A}}{1+\beta[A]/_{\alpha K_A}}\right) + [S_n]\left(\dfrac{1+\dfrac{[A]}{\alpha K_A}}{1+\beta[A]/_{\alpha K_A}}\right)}} \right) \frac{X}{dX/_{dS}.} \qquad (67)$$

From Eq. (67), let

$$\psi^1 = 1 + \frac{[A]}{\alpha K_A}. \qquad (68)$$

$$\psi^{11} = 1 + \frac{\beta[A]}{\alpha}. \qquad (69)$$

$$\psi^{111} = 1 + \frac{[A]}{K_A}. \qquad (70)$$

Substituting Eqs. (68), (69) and (70) into Eq. (67) yields

$$\frac{dS}{dt} = \left(\begin{array}{l} \mu_{\max}^{m} \dfrac{[S_1]}{K_{m1}+[S_1]} + \mu_{\max}^{m} \dfrac{[S_2]}{K_{m2}+[S_2]} + \mu_{\max}^{m} \dfrac{[S_3]}{K_{m3}+[S_3]} + \\ \mu_{\max}^{m} \dfrac{[S_4]}{K_{m4}+[S_4]} + \mu_{\max}^{m} \dfrac{[S_5]}{K_{m5}+[S_5]} \bullet \bullet \bullet + \mu_{\max}^{m} \dfrac{[S_n]}{K_{mn}+[S_n]} \end{array} \right) \frac{X}{dX/_{dS}.} \qquad (71)$$

Simplifying Eq. (71) yields

$$\frac{dS}{dt} = \left(\mu^m_{max} \frac{[S_1]}{K_{m1} + [S_1]} + \mu^m_{max} \frac{[S_2]}{K_{m2} + [S_2]} + \mu^m_{max} \frac{[S_3]}{K_{m3} + [S_3]} + \mu^m_{max} \frac{[S_4]}{K_{m4} + [S_4]} + \mu^m_{max} \frac{[S_5]}{K_{m5} + [S_5]} \bullet \bullet \bullet + \mu^m_{max} \frac{[S_n]}{K_{mn} + [S_n]} \right) X. \quad (72)$$

Similarly, rearranging Eq. (72) yields

$$\frac{dX}{X} = \left(\mu^m_{max_1} \frac{[S_1]}{K_{m1} + [S_1]} + \mu^m_{max_2} \frac{[S_2]}{K_{m2} + [S_2]} + \mu^m_{max_3} \frac{[S_3]}{K_{m3} + [S_3]} + \mu^m_{max_4} \frac{[S_4]}{K_{m4} + [S_4]} + \mu^m_{max_5} \frac{[S_5]}{K_{m5} + [S_5]} \bullet \bullet \bullet + \mu^m_{max_n} \frac{[S_n]}{K_{mn} + [S_n]} \right) dt. \quad (73)$$

Integrating Eq. (73) yields

$$In \frac{X}{X_o} = \left(\mu^m_{max_1} \frac{[S_1]}{K_{m1} + [S_1]} + \mu^m_{max_2} \frac{[S_2]}{K_{m2} + [S_2]} + \mu^m_{max_3} \frac{[S_3]}{K_{m3} + [S_3]} + \mu^m_{max_4} \frac{[S_4]}{K_{m4} + [S_4]} + \mu^m_{max_5} \frac{[S_5]}{K_{m5} + [S_5]} \bullet \bullet \bullet + \mu^m_{max_n} \frac{[S_n]}{K_{mn} + [S_n]} \right) t$$
$$+ C. \quad (74)$$

Where $C = $ constant. Finally, Eq. (74) can be written as

$$X = X_0 \ell^{\left(\mu^m_{max_1} \frac{[S_1]}{K_{m1} + [S_1]} + \mu^m_{max_2} \frac{[S_2]}{K_{m2} + [S_2]} + \mu^m_{max_3} \frac{[S_3]}{K_{m3} + [S_3]} \bullet \bullet \bullet + \mu^m_{max_n} \frac{[S_n]}{K_{mn} + [S_n]} \right)}. \quad (75)$$

$$\mu^m_{max} t = \sum_{1=i}^{n} \frac{K_{m_i} + [S_i]}{[S_i]} In \frac{X}{X_o} + C. \quad (76)$$

If $X_o = 1$ for ideal situation Eq. (76) becomes

$$\mu^m_{max} t = \sum_{1=i}^{n} \frac{K_{mi} + [S_i]}{[S_i]} In X + C. \quad (77)$$

In a multiple enzyme reaction the specific rate of biodegradation reaction is expressed as the sum for the individual enzymes. Hence, for a catalysed system we have

$$\mu_A^m = \left(\begin{array}{l} \mu_{A.max_1}^m \dfrac{[S_1]}{K_{m_1}\left(\dfrac{1+\dfrac{[A]}{K_A}}{1+\beta\dfrac{[A]}{\alpha K_A}}\right) + [S_1]\left(\dfrac{1+\dfrac{[A]}{K_A}}{1+\beta\dfrac{[A]}{\alpha K_A}}\right)} + \\[2em] \mu_{A.max_2}^m \dfrac{[S_2]}{K_{m_1}\left(\dfrac{1+\dfrac{[A]}{K_A}}{1+\beta\dfrac{[A]}{\alpha K_A}}\right) + [S_2]\left(\dfrac{1+\dfrac{[A]}{K_A}}{1+\beta\dfrac{[A]}{\alpha K_A}}\right)} \\[2em] + \bullet \bullet \bullet + \mu_{A.max_n}^m \dfrac{[S_n]}{K_{m_{n1}}\left(\dfrac{1+\dfrac{[A]}{K_A}}{1+\beta\dfrac{[A]}{\alpha K_A}}\right) + [S_n]\left(\dfrac{1+\dfrac{[A]}{K_A}}{1+\beta\dfrac{[A]}{\alpha K_A}}\right)} \end{array} \right) \dfrac{X}{dX/dS}$$

$$(78)$$

Substituting Eqs. (68), (67) and (68) into Eq. (78) yields;

$$\dfrac{dS}{dt} = \left(\begin{array}{l} \mu_{A.max_1}^m \dfrac{[S_1]}{K_{m_1}\left(\dfrac{\psi'''}{\psi''}\right)_1 + [S_n]\left(\dfrac{\psi'}{\psi''}\right)_1} + \mu_{A.max_n}^m \dfrac{[S_2]}{K_{m_n}\left(\dfrac{\psi'''}{\psi''}\right)_2 + [S_2]\left(\dfrac{\psi'}{\psi''}\right)_2} + \\[2em] \mu_{A.max_n}^m \dfrac{[S_n]}{K_{m_n}\left(\dfrac{\psi'''}{\psi''}\right)_n + [S_n]\left(\dfrac{\psi'}{\psi''}\right)_n} \end{array} \right) \dfrac{X}{dX/dS}.$$

$$(79)$$

Where ds/dt = velocity or specific rate. Equation (79) can be written in terms component i as

$$\dfrac{dS}{dt} = \sum_{i=1}^{n} \mu_{A.max_i}^m \dfrac{[S_i]}{K_{m_1}\left(\psi'''/_{\psi''}\right)_i + [S_i]\left(\dfrac{X}{dX/dS}\right)}.$$

$$(80)$$

where $\mu_{A.max}^m = \sum_{1=i}^{n} \mu_{A.max}^m$

Therefore

$$\mu_{A.\max}^{m} = \sum_{1=i}^{n} K_{m_i} \left(\psi''' \big/ \psi'' \right)_i + [S_i] \left(\frac{\left(\psi''' \big/ \psi'' \right)_i \frac{dX}{X}}{[S_i].} \right) \tag{81}$$

Integrating Eq. (81) yields

$$\mu_{A.\max}^{m} = \sum_{1=i}^{n} K_{m_i} \left(\psi''' \big/ \psi'' \right)_i + [S_i] \left(\frac{\left(\psi''' \big/ \psi'' \right)_i \frac{dX}{X}}{[S_i] \, In \, X + C_1.} \right) \tag{82}$$

2.2 Experimental Analysis

2.2.1 Physicochemical Analysis

Table 1.

2.2.2 Analysis for C-Group Hydrocarbon

The sample was collected in pond system located in Obite town in Ogba/Egbema/Ndoni Local Government Area of Nigeria.

The physical properties of the c-groups composition of hydrocarbon samples obtained from a pond in Obite town in Niger Delta Area of Nigeria was determined with the aid of gas chromatography (GC) made of AGILEND-HP68 90model and Varian 3900 were used for the same purpose. The photograph of the Gas chromatography (GC).

2.2.3 Pond Consideration for the Investigation

Figure 1 shows a schematic of the main oxidation pond system used in the experimental investigation of this thesis. Samples were collected at different surface distance, incline and vertical depth (subsurface) at every 4 weeks for laboratory analysis to be carried on the following parameters:

1. Microbial activity was determined at different points of the pond system to ascertain the effect of velocity due to continuous discharge of wastewater in a pond system.
2. Physicochemical parameters of the wastewater was determined at different points of the oxidation pond system to ascertain the distribution of biochemical oxygen demand (BOD), Chemical Oxygen Demand (COD) and dissolved oxygen concentration as a result of velocity effect.
3. Individual hydrocarbon concentration of the wastewater was determined at different points of pond system to ascertain the rate of degradation as a result of velocity transfer effect.

Table 1 Standard for measuring the physiochemical properties of some parameters present in waste water treatment

Parameters	Analytical method
General Appearance	APHA 2110
pH	APHA 4500H'8
Turbidity (NTU)	APHA 21308
Total Dissolved Solids (mg/l)	APHA 25108
Conductivity (μS/cm)	APHA 25110A
Biochemical Oxygen Demand (mg/l)	APHA 5210D
Chemical Oxygen Demand (mg/l)	APHA 5220D
Dissolved oxygen (mg/l)	APHA 5230D
Chloride (mg/l)	APHA 4500Cr8
Total Alkalinity (mg/l)	ASTM D10678
Nitrate (mg/l)	EPA 352.1
Phosphate (mg/l)	APHA 4500-PD
Sulphate (mg/l)	APHA 4500S0/"E
Total Iron (mg/l)	APHA 31118
Pseudomonas sp. (cfu/100 ml)	APHA 92228
Bacillus (cfu/100 ml)	APHA 92228
E-coli (cfu/100 ml)	ASTM D5392-93
Enterococci (cfu/100 ml)	ASTM D5295-92
Total Plate Count (cfu/ml)	APHA 9215C

Fig. 1 The main experimental investigation on the force acting on the immersed object-surface floater along side with the fluid element in a pond system (for continuous discharged of wastewater)

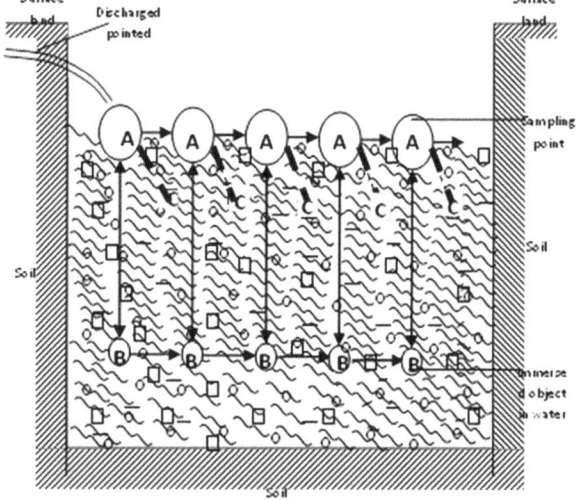

Before starting the experiments, the initial composition of the pond system was measured to determine the physicochemical parameters, microbial population and individual hydrocarbon concentration at different point in the pond system for dry seasons.

The wastewater contaminants velocity was determined at various sampling points A_1, A_2, A_3, A_4, C_1, C_2, C_3, C_4, B_1, B_2, B_3 and B_4. These points established because the sampling area considered during the investigation at various interval of days, the results obtained from the analysis was recorded. Since momentum is defined as a product of mass and its velocity. The specific rate (velocity) on the substrate concentration, microbial concentration and physiochemical parameters were measured at these various sampling points. Similarly the mass of the substrate, concentration, biomass concentration and as well as physicochemical parameter under investigation were determined experimentally. The experimental diagram for the investigation is illustrated in Fig. 1.

The diagram below shows the main experimental outline for the research.

Where, o is hydrocarbon, □ is particle, $\sim\sim$ is formation created as a result of disturbance (continuous discharge of waste water into the oxidation pond, $\rightarrow$ direction of flow A_0, A_1, A_2, A_3, A_4, B_0, B_1, B_2, B_3, B_4, C_0, C_1, C_2, C_3, and C_4 are sampling points.

Equation (82) is useful for predicting the performance of petroleum hydrocarbon degraders and the evaluation of the specific and maximum specific growth rates in the biodegradation reaction of petroleum hydrocarbon for single and multiple enzymes under the influence of velocity distribution.

3 Results and Discussion

The result obtained from the investigation is presented in Tables and Figures as shown in this paper. The result revealed the effect of velocity distribution on substrate concentration biomass built up and physiochemical parameters concentration.

Where $V_I^2 + 0.14V_I - 0.14 = 0$. Figure 2: Mathematical expression into the equation of intercept yield a positive and negative velocity which was classified as the inflow velocity (positive) and backflow velocity (negative).

The intercept on the graph of $V_I^2/1 - V_I$ against L was established as $V_I^2/1 - V_I$ whereas the $gs_o = slope$ was given as $S_0 = 0.625 \times 10^{-4}$. Therefore $gs_o = 5.625 \times 9.81$ as presented.

The result presented in Fig. 3 illustrates the relationship between $1_{\mu E}$ and T/S using the application of Lineweaver Burk plot. The best fit line was established to evaluate the $\mu^E = 10.0 \times 10^{-5}(day^{-1})$ $K_m^E = 200$ as presented in Table 7. The variation on the functional coefficient can be attributed to the variation in velocity distribution, biomass built up, physiochemical properties of the pond system.

Figure 4 illustrates the relationship between n the theoretical evaluation of $^1/_{\mu^T}$ and $^{1/s \times 10^{-2}}$ using the application of the Lineweaver Bulk plot. The best line of the fit was established to determine the coefficient values of $\mu_{max}^T = 45.45 \times 10^{-4}(day)^{-1}$ $K_{max}^T = 0.0125$. The variation on the functional coefficient can be attributed to the variation in velocity distribution biomass built up, physiochemical properties of the pond system.

The mathematical computation of some functional parameters for dry season was presented in Table 2. The velocity, gravity and surface slop upon the influence of velocity was determined. The variation in each of the functional parameter can be attributed to the variation in the distance as shown in Table 2. The result presented in Table 3 illustrates the specific rate of microbial growth on the pond

Fig. 2 $V_I^2/1 - V_I$ versus L

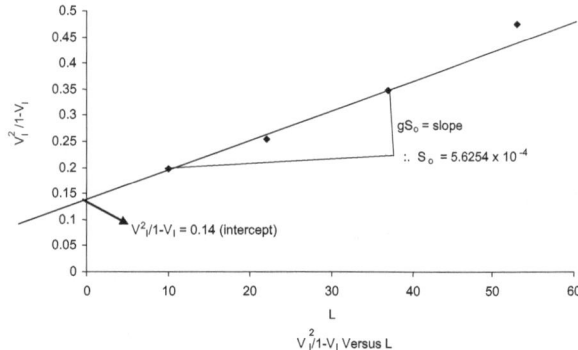

Fig. 3 Dry season
Lineweaver Bulk plot for
1/u versus 1/s

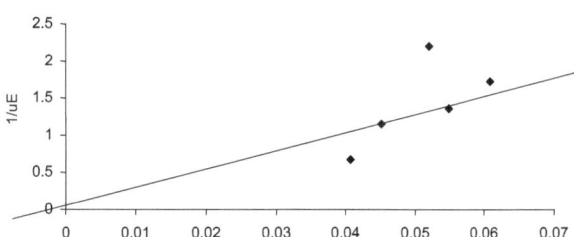

Fig. 4 Dry season
Lineweaver Burk plot for
1/U versus 1/S

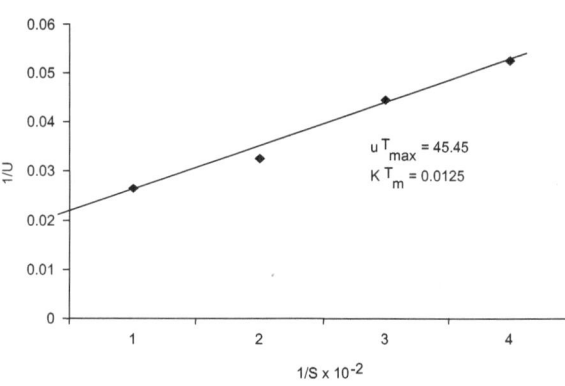

Table 2 Mathematical computation of some functional parameters for dry season

Distance (m)	V_I (m/day)	V_I^2	$\frac{V_I^2}{1-V_I}$	g (m/s²)	S_o
10	0.357	0.127	0.198	9.81	5.6254×10^{-4}
22	0.393	0.154	0.254		
37	0.441	0.194	0.347		
53	0.474	0.225	0.475		

Table 3 Specific rate of microbial growth on the pond using the theoretical model developed (dry season)

Time (weeks)	Distance (m)	Substrate conc. Mol % [S]	μ^T	1/[S]	$1/\mu^T$	$\Delta\mu\% = \dfrac{\mu^T - \mu^E}{\mu^E} \times 100$	$\overline{\Delta\mu} = \dfrac{1}{n}\sum_1^n \Delta\mu_i$
0	0	45.061	–	0.0222	–	–	–
4	10	37.650	7.50×10^3	0.0266	1.333×10^{-4}	−91.38	-0.18
8	22	30.807	3.97×10^4	0.0325	2.64×10^{-5}	−16.15	−0.0323
12	37	22.499	1.92×10^5	0.0444	5.21×10^{-6}	160.87	0.32174
16	53	19.012	9.73×10^5	0.0526	1.03×10^{-6}	1,583.39	3.167

$\mu_{\text{max}}^T = 45.45 \times 10^{-4} (\text{day})^{-1}$ and $K_{\text{m}}^T = 0.0125$

Table 4 Microbial population on the pond system for dry season analysis before receiving 2 months wastewater discharged (dry season)

| Time (weeks) | Distance (m) | Microbial population (cfu/ml) | | | |
		E. coli	Enterococci organisms	Pseudomonas sp.	Bacillus
0	0	2.2×10^2	3.2×10^2	9.0×10^2	4.2×10^1
4	10	3.5×10^4	4.8×10^4	2.5×10^3	1.6×10^3
8	22	1.7×10^4	2.5×10^3	2.1×10^3	1.1×10^3
12	37	3.0×10^4	3.8×10^4	3.6×10^3	2.0×10^3
16	53	2.3×10^4	3.0×10^4	3.3×10^3	1.5×10^3

Table 5 Specific rate of microbial growth on the pond using the experimental data (dry season)

Time (weeks)	Distance (m)	Substrate concentration Mol % [S]	μ^E	1/[S]	$1/\mu^E$
0	0	45.061	–	0.0222	–
4	10	37.650	8.71×10^4	0.0266	1.148×10^{-5}
8	22	30.807	4.52×10^4	0.0325	2.212×10^{-5}
12	37	22.499	7.36×10^4	0.0444	1.359×10^{-5}
16	53	19.012	5.78×10^4	0.0526	1.730×10^{-5}

$\mu^E_{max} = 10.0 \times 10^{-5} (\text{day})^{-1}$; $K^E_{max} = 200$

Table 6 Microbial population on the pond system for dry season analysis before receiving more wastewater discharged (dry season)

| Time (weeks) | Distance (m) | Microbial population (cfu/ml) | | | |
		E. coli	Enterococci organisms	Pseudomonas sp.	Bacillus
0	0	2.6×10^1	1.0×10^1	2.5×10^2	1.3×10^2
4	10	1.3×10^4	5.3×10^3	8.0×10^4	2.7×10^4
8	22	3.8×10^4	2.2×10^4	4.1×10^5	5.0×10^4
12	37	4.1×10^4	2.8×10^4	6.3×10^6	7.1×10^4
16	53	2.0×10^4	2.6×10^4	3.7×10^4	2.8×10^4

using the theoretical model developed for the dry season model. The relationship between distance and line per unit change was considered which leads to constant monitoring of the substrate and establishment of the specific rate as well as the maximum specific rate as presented in Table 4. Table 4 illustrates the microbial population or biomass built up in pond system upon the influence of velocity distribution, lag progressive stationary and decline phase was observed as a result of velocity distribution effect on the pond system (Tables 5 and 6). Table 7 Illustrate the inflow velocity distribution as well as backflow distribution of the contaminants in the pond system which was evaluated to determine the actual velocity of distribution of contaminants in the medium. The variation of obtained in the actual velocity can be attributed to the variation in the distance positive and negative velocity as well as the concentration of the contaminant. From Table 8 it is seen that the mean physiochemical properties of the pond system changes with time as well distance. This indicates the effect of velocity distribution on the physiochemical

Table 7 Dry season velocity for positive and negative velocity in pond system

| Distance | Velocity | | Actual velocity (AV$_{L1}$) in terms of distance (m/day) actual velocity AV = (+V) − (−V) |
	+V$_{L1}$ (m/day)	−V$_{L1}$ (m/day)	
0	5.115	1.142	3.973
10	3.449	0.273	3.176
22	2.149	0.170	1.179
37	1.190	0.0938	1.181
53	0.634	0.0499	0.629

Table 8 Experimentally determined mean physiochemical properties of some wastewater discharged with the pond system at dry season

| Parameter | | Mean concentration of the parameters | | | | | | | |
Sampling (week)	Unit	Week 2	Week 4	Week 6	Week 8	Week 10	Week 12	Week 14	Week 16
pH at recorded temp.		6.78	6.96	6.83	6.88	7.14	7.115	6.87	7.00
Measured temp of pH	°C	28	27.2	28	28.4	27.6	27.7	26.9	28.3
Turbidity	NTu	30	24	32	23	27	22	20	28
Conductivity	mS/cm	11.7	12.6	12.8	11.5	14.5	13.7	11.7	12.0
Alkalinity/ Bicarbonate	mg/l	50.8	51	25.5	27	78.7	72.6	42.0	56.0
Total suspended solid	mg/l	42.00	56.10	47.50	764.01	10.33	408.13	60.3	120.0
Total dissolved solid	mg/l	6.215	6,486	7,245	7118	7,015	7,050	7,121	6,405
BOD (biochemical oxygen demand)	mg/l	20	8.4	25	14	17.4	13.5	14.1	9.0
COD (Chemical oxygen demand)	mg/l	59.5	48	40.1	50.2	55	60.2	45.2	50.0
Salinity (as chloride)	mg/l	10,700	10,645	7,526	8,236	1,500	11,652	8,540	7,770
Iron	mg/l	0.51	0.32	0.1	0.12	0.15	0.18	0.11	0.20
Nitrate	mg/l	4	2.5	5.5	3.2	1.8	1.7	2.7	1.3
Sulphate	mg/l	1,550	1,500	1,400	1,340	1,820	1,855	1,280	1,560
DO (Dissolved oxygen)	mg/l	3.98	3.92	3.72	4.9	4	1.7	2.96	3.34

properties of the pond as well as it effect on the biomass built up which in general influence the hydrocarbon degradation.

The effect of velocity distribution on the hydrocarbon was examined as presented in Table 9. The velocity distribution influence the substrate concentration either increasing or decreasing order as well examined experimental, which is presented in Tables 9 and 10.

Table 9 Experimentally determined the specific rate (Velocity) substrate concentrations at various surface distance, inclined depths and in terms of vertical depth in the pond system for dry season

Time (week)	Distance (m)	Velocity (m/weak)	Component	Substrate concentration at various surface distance (mol %)	Vertical and inclined depth (m)	Substrate concentration at the vertical depth (mol %)					Substrate concentration at inclined depth (mol %)				
						0.970	0.501	0.351	0.090	0.007	0.970	0.501	0.351	0.090	0.007
0	0	0	i-C$_5$	0.970	0	0.970	0.501	0.351	0.090	0.007	0.970	0.501	0.351	0.090	0.007
			n-C$_5$	1.364		1.364	1.036	0.526	0.136	0.008	1.364	1.036	0.526	0.136	0.008
			C$_6$	3.831		3.831	2.761	1.645	0.670	0.052	3.831	2.761	1.645	0.670	0.052
			C$_7$	1.486		1.486	1.034	0.558	0.290	0.152	1.486	1.034	0.558	0.290	0.152
			C$_8$	6.035		6.035	5.076	3.436	1.586	1.150	6.035	5.076	3.436	1.586	1.150
			C$_9$	9.833		9.833	6.238	4.673	2.613	1.908	9.833	6.238	4.673	2.613	1.908
			C$_{10}$	9.042		9.042	8.628	8.013	7.513	7.199	9.042	8.628	8.013	7.513	7.199
			C$_{11}$	12.50		12.50	12.38	11.64	9.597	8.000	12.50	12.38	11.64	9.597	8.000
4	10	2.5	i-C$_5$	0.501	0.3	0.060	0.053	0.032	0.003	0.001	0.060	0.053	0.031	0.003	0.001
			n-C$_5$	1.036		0.080	0.071	0.064	0.018	0.005	0.070	0.070	0.064	0.017	0.003
			C$_6$	2.761		0.013	0.013	0.010	0.010	0.004	0.012	0.012	0.010	0.10	0.002
			C$_7$	1.034		0.052	0.042	0.027	0.016	0.005	0.052	0.040	0.028	0.017	0.002
			C$_8$	5.076		0.078	0.051	0.042	0.009	0.003	0.064	0.53	0.044	0.009	0.001
			C$_9$	6.238		0.086	0.036	0.015	0.008	0.002	0.077	0.038	0.017	0.008	0.002
			C$_{10}$	8.628		0.381	0.202	0.096	0.033	0.009	0.383	0.202	0.094	0.036	0.008
			C$_{11}$	12.376		0.407	0.258	0.053	0.024	0.008	0.410	0.257	0.053	0.023	0.007
8	22	2.75	i-C$_5$	0.351	0.6	0.041	0.030	0.020	0.008	0.002	0.044	0.031	0.020	0.008	0.002
			n-C$_5$	0.526		0.050	0.006	0.023	0.002	0.001	0.049	0.006	0.003	0.002	0.001
			C$_6$	1.645		0.010	0.008	0.002	0.002	0.001	0.010	0.007	0.002	0.002	0.001
			C$_7$	0.558		0.031	0.025	0.004	0.008	0.002	0.028	0.023	0.001	0.007	0.002
			C$_8$	3.436		0.038	0.021	0.008	0.008	0.001	0.038	0.020	0.001	0.006	0.001
			C$_9$	4.673		0.050	0.035	0.024	0.010	0.005	0.047	0.035	0.022	0.010	0.006
			C$_{10}$	8.013		0.201	0.166	0.007	0.026	0.007	0.213	0.167	0.091	0.027	0.005
			C$_{11}$	11.641		0.200	0.141	0.085	0.029	0.008	0.206	0.191	0.089	0.030	0.006

Table 10 Experimentally determined the specific rate (Velocity) substrate concentrations at various surface distance, inclined depths and in terms of vertical depth in the pond system for dry season

Time (week)	Distance (m)	Velocity (m/weak)	Component	Substrate concentration at various surface distance (mol %)	Vertical and Inclined depth (m)	Substrate concentration at the vertical depth (mol %)	Substrate concentration at inclined depth (mol %)
12	37	3.08	i-C$_5$	0.090	0.9	0.030 0.004 0.001 0.000 0.000	0.020 0.004 0.002 0.000 0.000
			n-C$_5$	0.136		0.022 0.003 0.003 0.000 0.000	0.027 0.002 0.002 0.000 0.000
			C$_6$	0.670		0.006 0.004 0.002 0.000 0.000	0.063 0.004 0.003 0.000 0.000
			C$_7$	0.290		0.10 0.005 0.001 0.002 0.000	0.011 0.003 0.002 0.000 0.000
			C$_8$	1.586		0.026 0.009 0.003 0.006 0.001	0.015 0.008 0.003 0.002 0.000
			C$_9$	2.613		0.031 0.028 0.008 0.007 0.002	0.023 0.028 0.007 0.006 0.001
			C$_{10}$	7.513		0.170 0.061 0.009 0.009 0.003	0.201 0.062 0.007 0.007 0.002
			C$_{11}$	9.597		0.103 0.089 0.015 0.032 0.005	0.184 0.089 0.016 0.009 0.004
16	53	3.31	i-C$_5$	0.007	1.2	0.010 0.001 0.00 0.000 0.000	0.006 0.001 0.000 0.000 0.000
			n-C$_5$	0.0079		0.011 0.001 0.00 0.000 0.000	0.008 0.000 0.000 0.000 0.000
			C$_6$	0.0517		0.003 0.00 0.00 0.000 0.000	0.004 0.000 0.000 0.000 0.000
			C$_7$	0.152		0.009 0.001 0.00 0.000 0.000	0.005 0.001 0.000 0.000 0.000
			C$_8$	1.150		0.013 0.001 0.00 0.000 0.000	0.005 0.002 0.000 0.000 0.000
			C$_9$	1.908		0.101 0.002 0.001 0.002 0.001	0.061 0.004 0.001 0.001 0.001
			C$_{10}$	7.199		0.133 0.005 0.001 0.001 0.001	0.092 0.004 0.001 0.001 0.001
			C$_{11}$	8.000		0.096 0.003 0.001 0.002 0.001	0.137 0.005 0.001 0.001 0.002

Nomenclature

$[A]$	Activator
$[EA]$	Enzyme activator
$[EAI]$	Enzyme activator inhibitor
$[EAS]$	Enzyme activator substrate
$[EI]$	Enzyme inhibitor
$[ES]$	Enzyme activator
$C_1\,C_2$	Constant
Cn	Coefficient of momentum transfer
E	Enzyme
g	Acceleration due to gravity (m/s^2)
K_{-1}, k_1	Equilibrium constant for forward and backward reactor
K_A	Equilibrium constant of activator
Kp	Equilibrium constant of the product
Ks	Equilibrium constant of substrate
L, x	distance (m)
n	Number of cells in the reactor volume $(ycell/cm^2)$
R_{max}	The maximum degradation rate of C-group hydrocarbon (mg/l-d)
S	Liquid concentration of petroleum hydrocarbon (mg/Hc/l)
$\lfloor S \rfloor$	Substrate concentration (mol %)
S_o	Below the surface water slope
T, t	Time (s)
V	Velocity (m/s^2)
X	Biomass concentration (cfu/ml)
y	The cell number produced per unit amount of substrate consumed (gcell/mol %)
α	Constant
β	Constant
λ^2	Constant
μ	Momentum transfer (g m/s)
μ^S	Specific growth rate of single enzyme
μ_{max}^T	Maximum growth rate (mg/l/day) for theoretical
μ_{max}^E	Maximum growth rate (mg/l/day) for experimental
μ_{max}^S	Maximum growth rate for single enzyme (mg/l/day)
Ψ', Ψ'', Ψ'''	Coefficient of expression of constant

References

1. E. Arvin, B.K. Jensen, A.T. Gundersen, Wat. Sci. Tech. **23**(7), 1375–1384 (1991)
2. B. Polec, Gaz. Cukro. **99**(1), 17–20 (1991)
3. A. Ogoni, J. Model. Simulat. Contr. **62**(4), 11–20 (2001)
4. M. Abowei, E. Wami, J. Model. Simulat. Contr. AMSE **15**(4), 1–17 (1988)
5. D. Goldsmith, R. Balderson, Hazard. Waste Hazard. Mater. **6**(2), 145–154 (1989)
6. P. Diagrazia, J. Blackburn, B. Bienkowski, B. Hilton, G. Reed, J. King, G. Sayler, Appl. Biochem. Biotech. **24/25**(12), 237–252 (1990)
7. T. Premuzic, M. Lin, SPE **2048**, 1693–1790 (1990)
8. M. Miller, M. Alexander, Environ. Sci. Tech. **25**(2), 250–245 (1991)
9. P. Alvarez, T. Vogel, Biodegradation **21**, 43–51 (1991)
10. M. Bradley, F. Chapelle, Environ. Sci. Tech. **32**, 553–557 (1996)
11. D. Bryniok, P. Koziollek, S. Baner, H. Knackmuss, (Battelle, Columbus, OH, 1998), pp. 181–186
12. A. Ogulu, V. Omubo-Pepple, J. Model. Simulat. Contr. **62**(4), 43–54 (2000)
13. O. Odigure, A. Abdulkareem, J. Model. Simulat. Contr. **62**(3), 57–63 (2001)
14. A. Ahmed Melegy, A. Mohamed, A. Gamal, J. Model. Simulat. Contr. **63**(3), 30–36 (2003)
15. A. Ahmed Melegy, J. Model. Simulat. Contr. **65**(1), 35–48 (2004)
16. U. Ubaezue, K. Egereonu, J. Model. Simulat. Contr. **65**(1), 59–76 (2004)
17. Hong-gyu, B. Richard, Appl. Environ. Microbiol. **14**, 72–76 (1990)
18. C. Ukpaka, PhD, research work, Rivers State University of Science and Technology, 2009, pp. 210–300
19. E. Hinchee, S. Ong, R. Miller, D. Downey, R. Frandl, USaF, 101 (1992)
20. J. Dibble, R. Bartha, Appl. Environ. Microbiol. **37**(4), 729–739 (1979)
21. S. Khan, J. Model. Simulat. Contr. **61**(2), 30–40 (2000)
22. M.M. Grady, I.P. Wright, P.K. Swart, C.T. Pillinger, The carbon and oxygen isotopic composition of meteoritic carbonates. Geochim. Cosmochim. Acta **52**(12), 2855–2866 (1988)